BIOLOGY
AND ITS MAKERS

With Portraits and Other Illustrations

BY
WILLIAM A. LOCY, PH.D., SC.D.
Professor in Northwestern University

NEW YORK
HENRY HOLT AND COMPANY
1908

ILLUSTRATED &
PUBLISHED BY
E-KİTAP PROJESİ & CHEAPEST BOOKS

www.facebook.com/EKitapProjesi

www.cheapestboooks.com

Istanbul
ISBN:
978-625-6629-81-3

Published June, 1908

Cover Image: The Scheleton from VESALIUS
"Plate from Vesalius's De Fabrica"

To

MY GRADUATE STUDENTS

Who have worked by my side in the Laboratory

Inspired by the belief that those who seek shall find

This account of the findings of some of

The great men of biological science

Is dedicated by

THE AUTHOR

PREFACE

The writer is annually in receipt of letters from students, teachers, ministers, medical men, and others, asking for information on topics in general biology, and for references to the best reading on that subject. The increasing frequency of such inquiries, and the wide range of topics covered, have created the impression that an untechnical account of the rise and progress of biology would be of interest to a considerable audience. As might be surmised, the references most commonly asked for are those relating to different phases of the Evolution Theory; but the fact is usually overlooked by the inquirers that some knowledge of other features of biological research is essential even to an intelligent comprehension of that theory.

In this sketch I have attempted to bring under one view the broad features of biological progress, and to increase the human interest by writing the story around the lives of the great Leaders. The practical execution of the task resolved itself largely into the question of what to omit. The number of detailed researches upon which progress in biology rests made rigid selection necessary, and the difficulties of separating the essential from the less important, and of distinguishing between men of temporary notoriety and those of enduring fame, have given rise to no small perplexities.

The aim has been kept in mind to give a picture sufficiently diagrammatic not to confuse the general reader, and it is hoped that the omissions which have seemed necessary will, in a measure, be compensated for by the clearness of the picture. References to selected books and articles have been given at the close of the volume, that will enable readers who wish fuller information to go to the best sources.

The book is divided into two sections. In the first are considered the sources of the ideas—except those of organic evolution—that dominate biology, and the steps by which they have been molded into a unified science. The Doctrine of Organic Evolution, on account of its importance, is reserved for special consideration in the second section. This is, of course, merely a division of convenience, since after its acceptance the doctrine of evolution has entered into all phases of biological progress.

The portraits with which the text is illustrated embrace those of nearly all the founders of biology. Some of the rarer ones are unfamiliar even to biologists, and have been discovered only after long search in the libraries of Europe and America.

An orderly account of the rise of biology can hardly fail to be of service to the class of inquirers mentioned in the opening paragraph. It is hoped that this sketch will also meet some of the needs of the increasing body of students who are doing practical work in biological laboratories. It is important that such students, in addition to the usual class-room instruction, should get a perspective view of the way in which biological science has come into its present form.

The chief purpose of the book will have been met if I have succeeded in indicating the sources of biological ideas and the main currents along which they have advanced, and if I have succeeded, furthermore, in making readers acquainted with those men of noble purpose whose work has created the epochs of biological history, and in showing that there has been continuity of development in biological thought.

Of biologists who may examine this work with a critical purpose, I beg that they will think of it merely as an outline sketch which does not pretend to give a complete history of biological thought. The story has been developed almost entirely from the side of animal life; not that the botanical side has been underestimated, but that the story can be told from either side, and my first-hand acquaintance with botanical investigation is not sufficient to justify an attempt to estimate its particular achievements.

The writer is keenly aware of the many imperfections in the book. It is inevitable that biologists with interests in special fields will miss familiar names and the mention of special pieces of notable work, but I am drawn to think that such omissions will be viewed leniently, by the consideration that those best able to judge the shortcomings of this sketch will also best understand the difficulties involved.

The author wishes to acknowledge his indebtedness to several publishing houses and to individuals for permission to copy cuts and for assistance in obtaining portraits. He takes this opportunity to express his best thanks for these courtesies. The parties referred to are the director of the American Museum of Natural History; D. Appleton & Co.· P. Blakiston's Sons & Co.; The Macmillan Company; The Open Court Publishing Company; the editor of the *Popular Science Monthly*; Charles Scribner's Sons; Professors Bateson, of Cambridge, England; Conklin, of Philadelphia; Joubin, of Rennes, France; Nierstrasz, of Utrecht, Holland; Newcombe, of Ann Arbor, Michigan; Wheeler and E. B. Wilson, of New York City. The editor of the *Popular Science Monthly* has also given permission to reprint the substance of Chapters IV and X, which originally appeared in that publication.

W. A. L.

NORTHWESTERN UNIVERSITY,
Evanston, Ill., April, 1908.

CONTENTS

PART I

The Sources of Biological Ideas Except Those of Organic Evolution

CHAPTER I

PART II

The Doctrine of Organic Evolution

CHAPTER XVI

CHAPTER XVII

ILLUSTRATIONS

PART I

THE SOURCES OF BIOLOGICAL IDEAS EXCEPT THOSE OF ORGANIC EVOLUTION

CHAPTER I

AN OUTLINE OF THE RISE OF BIOLOGY AND OF THE EPOCHS IN ITS HISTORY

"Truth is the Daughter of Time."

THE nineteenth century will be for all time memorable for the great extension of the knowledge of organic nature. It was then that the results of the earlier efforts of mankind to interpret the mysteries of nature began to be fruitful; observers of organic nature began to see more deeply into the province of life, and, above all, began to see how to direct their future studies. It was in that century that the use of the microscope made known the similarity in cellular construction of all organized beings; that the substance, protoplasm, began to be recognized as the physical basis of life and the seat of all vital activities; then, most contagious diseases were traced to microscopic organisms, and as a consequence, medicine and surgery were reformed; then the belief in the spontaneous origin of life under present conditions was given up; and it was in that century that the doctrine of organic evolution gained general acceptance. These and other advances less generally known created an atmosphere in which biology—the great life-science—grew rapidly.

In the same period also the remains of ancient life, long since extinct, and for countless ages embedded in the rocks, were brought to light, and their investigation assisted materially in understanding the living forms and in tracing their genealogy.

As a result of these advances, animal organization began to have a different meaning to the more discerning naturalists, those whose discoveries began to influence the trend of thought, and finally, the idea which had been so often previously expressed became a settled conviction, that all the higher forms of life are derived from simpler ones bv a gradual process of modification.

Besides great progress in biology, the nineteenth century was remarkable for similar advances in physics and chemistry. Although these subjects purport to deal with inorganic or lifeless nature, they touch biology in an intimate way. The vital processes which take place in all animals and plants have been shown to be physico-chemical, and, as a consequence, one must go to both physics and chemistry in order to understand them. The study of organic chemistry in late years has greatly influenced biology; not only have living products been analyzed, but some of them have already been constructed in the chemical laboratory. The formation of living matter through chemical means is still far from the thought of most chemists, but very complex organic compounds, which were formerly known only as the result of the action of life, have been produced, and the possibilities of further advances in that direction are very alluring. It thus appears that the discoveries in various fields have worked together for a better comprehension of nature.

The Domain of Biology.—The historv of the transformation of opinion in reference to living organisms is an interesting part of the story of intellectual development. The central subject that embraces it all is biology. This is one of the fundamental sciences, since it embraces all questions relating to life in its different phases and manifestations. Everything pertaining to the structure, the development, and the evolution of living organisms, as well as to their physiology, belongs to biology. It is now of commanding importance in the world of science, and it is coming more and more to be recognized that it occupies a field of compelling interest not only for medical men and scholars, but for all intelligent people. The discoveries and conquests of biology have wrought such a revolution in thought that they should be known to all persons of liberal culture. In addition to making acquaintance with the discoveries, one ought to learn something about the history of biology; for it is essential to know how it took its rise, in order to understand its present position and the nature of its influence upon expanding ideas regarding the world in which we live.

In its modern sense, biology did not arise until about 1860, when the nature of protoplasm was first clearly pointed out by Max Schultze, but the currents that united to form it had long been flowing, and we can never understand the subject without going back to its iatric condition, when what is now biology was in the germ and united with medicine. Its separation from medicine, and its rise as an independent subject, was owing to the steady growth of that zest for exploration into unknown fields which began with the new birth of science in the sixteenth century, and has continued in fuller measure to the present. It was the outcome of applying observation and experiment to the winning of new truths.

Difficulties.—But biology is so comprehensive a field, and involves so many details, that it is fair to inquire: can its progress be made clear to the reader who is unacquainted with it as a laboratory study? The matter will be simplified by two general observations—first, that the growth of biology is owing to concurrent progress in three fields of research, concerned, respectively, with the structure or architecture of living beings, their development, and their physiology. We recognize also a parallel advance in the systematic classification of animals and plants, and we note, furthermore, that the idea of evolution permeates the whole. It will be necessary to consider the advances in these fields separately, and to indicate the union of the results into the main channel of progress. Secondly, in attempting to trace the growth of ideas in this department of learning one sees that there has been a continuity of development. The growth of these notions has not been that of a chaotic assemblage of ideas, but a well-connected story in which the new is built upon the old in orderly succession. The old ideas have not been completely superseded by the new, but they have been molded into new forms to keep pace with the advance of investigation. In its early phases, the growth of biology was slow and discursive, but from the time of Linnæus to Darwin, although the details were greatly multiplied, there has been a relatively simple and orderly progress.

Facts and Ideas.—There are many books about biology, with directions for laboratory observation and experiment, and also many of the leading facts of the science have been given to the public, but an account of the growth of the ideas, which are interpretations of the facts, has been rarely attempted. From the books referred to, it is almost impossible to get an idea of biology as a unit; this even the students in our universities acquire only through a coherent presentation of the subject in the classroom, on the basis of their work in the laboratory. The critical training in the laboratory is most important, but, after all, it is only a part, although an essential part, of a knowledge of biology. In general, too little attention is paid to interpretations and the drill is confined to a few facts. Now, the facts are related to the ideas of the science as statistics to history—meaningless without interpretation. In the rise of biology the facts have accumulated constantly, through observation and experiment, but the general truths have emerged slowly and periodically, whenever there has been granted to some mind an insight into the meaning of the facts. The detached facts are sometimes tedious, the interpretations alwavs interesting.

The growth of the knowledge of organic nature is a long story, full of human interest. Nature has been always the same, but the capacity of man as its interpreter has varied. He has had to pass through other forms of intellectual activity, and gradually to conquer other phases of natural phenomena, before entering upon that most difficult task of investigating the manifestations of life. It will be readily understood, therefore, that biology was delayed in its development until after considerable progress had been made in other sciences.

It is an old saying that "Truth is the daughter of Time," and no better illustration of it can be given than the long upward struggle to establish even the elemental truths of nature. It took centuries to arrive at the conception of the uniformity of nature, and to reach any of those generalizations which are vaguelv spoken of as the laws of nature.

The Men of Science.—In the progress of science there is an army of observers and experimenters each contributing his share, but the rank and file supply mainly isolated facts, while the ideas take birth in the minds of a few gifted leaders, either endowed with unusual insight, or so favored bv circumstances that they reach general conclusions of importance. These advance-guards of intellectual conquest we designate as founders. What were they like in appearance? Under what conditions did they work, and what was their chief aim? These are interesting questions which will receive attention as our narrative proceeds.

A study of the lives of the founders shows that the scientific mood is pre-eminently one of sincerity. The men who have added to the growth of science were animated by an unselfish devotion to truth, and their lasting influence has been in large measure a reflection of their individual characters. Only those have produced permanent results who have interrogated nature in the spirit of devotion to truth and waited patiently for her replies. The work founded on selfish motives and vanity has sooner or later fallen by the wayside. We can recognize now that the work of scientific investigation, subjected to so much hostile criticism as it appeared from time to time, was undertaken in a reverent spirit, and was not iconoclastic, but remodelling in its influence. Some of the glories of our race are exhibited in the lives of the pioneers in scientific progress, in their struggles to establish some great truth and to maintain intellectual integrity.

The names of some of the men of biology, such as Harvey, Linnæus, Cuvier, Darwin, Huxley, and Pasteur, are widely known because their work came before the people, but others equally deserving of fame on account of their contributions to scientific progress will require an introduction to most of our readers.

In recounting the story of the rise of biology, we shall have occasion to make the acquaintance of this goodly company. Before beginning the narrative in detail, however, we shall look summarily at some general features of scientific progress and at the epochs of biology.

The Conditions under which Science Developed

In a brief sketch of biology there is relatively little in the ancient world that requires notice except the work of Aristotle and Galen; but with the advent of Vesalius, in 1543, our interest begins to freshen, and, thereafter, through lean times and fat times there is always something to command our attention.

The early conditions must be dealt with in order to appreciate what followed. We are to recollect that in the ancient world there was no science of biology as such; nevertheless, the germ of it was contained in the medicine and the natural history of those times.

There is one matter upon which we should be clear: in the time of Aristotle nature was studied by observation and experiment. This is the foundation of all scientific advancement. Had conditions remained unchanged, there is reason to believe that science would have developed steadily on the basis of the Greek foundation, but circumstances, to be spoken of later, arose which led not only to the complete arrest of inquiry, but also, the mind of man being turned away from nature, to the decay of science.

Aristotle the Founder of Natural History.—The Greeks represented the fullest measure of culture in the ancient world, and, naturally, we find among them the best-developed science. All the knowledge of natural phenomena centered in Aristotle (384–322 B.C.), and for twenty centuries he represented the highest level which that kind of knowledge had attained.

It is uncertain how long it took the ancient observers to lift science to the level which it had at the beginning of Aristotle's period, but it is obvious that he must have had a long line of predecessors, who had accumulated facts of observation and had molded them into a system before he perfected and developed that system. We are reminded that all things are relative when we find Aristotle referring to the ancients; and well he might, for we have indubitable evidence that much of the scientific work of antiquity has been lost. One of the most striking discoveries pointing in that direction is the now famous papyrus which was found by Georg Ebers in Egypt about 1860. The recent translation of this ancient document shows that it was a treatise on medicine, dating from the fifteenth century B.C. At this time the science of medicine had attained an astonishingly high grade of development among that people. And since it is safe to assume that the formulation of a system of medicine in the early days of mankind required centuries of observation and practice, it becomes apparent that the manuscript in question was no vague, first attempt at reducing medicine to a system. It is built upon much scientific knowledge, and must have been preceded by writings both on medicine and on its allied sciences.

It is not necessary that we should attempt to picture the crude beginnings of the observation of animated nature and the dawning of ideas relative to animals and plants; it is suitable to our purpose to commence with Aristotle, and to designate him, in a relative sense, as the founder of natural history.

That he was altogether dissatisfied with the state of knowledge in his time and that he had high ideals of the dignity of science is evidenced in his writings. Although he refers to the views of the ancients, he regarded himself in a sense as a pioneer. "I found no basis prepared," he says, "no models to copy. . . . Mine is the first step, and therefore a small one, though worked out with much thought and hard labor. It must be looked at as a first step and judged with indulgence." (From Osborn's *From the Greeks to Darwin.*)

There is general agreement that Aristotle was a man of vast intellect and that he was one of the greatest philosophers of the ancient world. He has had his detractors as well as his partisan adherents. Perhaps the just estimate of his attainments and his position in the history of science is between the enthusiastic appreciation of Cuvier and the critical estimate of Lewes.

This great man was born in Stagira in the year 384 B.C., and lived until 322 B.C. He is to be remembered as the most distinguished pupil of Plato, and as the instructor of

Alexander the Great. Like other scholars of his time, he covered a wide range of subjects; we have mention, indeed, of about three hundred works of his composition, many of which are lost. He wrote on philosophy, metaphysics, psychology, politics, rhetoric, etc., but it was in the domain of natural history that he attained absolute pre-eminence.

His Position in the Development of Science.—It is manifestly unjust to measure Aristotle by present standards; we must keep always in mind that he was a pioneer, and that he lived in an early day of science, when errors and crudities were to be expected. His greatest claim to eminence in the history of science is that he conceived the things of importance and that he adopted the right method in trying to advance the knowledge of the natural universe. In his program of studies he says: "First we must understand the phenomena of animals; then assign their causes; and, finally, speak of their generation." His position in natural history is frequently misunderstood. One of the most recent writers on the history of science, Henry Smith Williams, pictures him entirely as a great classifier, and as the founder of systematic zoölogy. While it is true that he was the founder of systematic zoölogy, as such he did not do his greatest service to natural history, nor does the disposition to classify represent his dominant activity. In all his work classification is made incidental and subservient to more important considerations. His observations upon structure and development, and his anticipation of the idea of organic evolution, are the ones upon which his great fame rests. He is not to be remembered as a man of the type of Linnæus; rather is he the forerunner of those men who looked deeper than Linnæus into the structure and development of animal life—the morphologists.

Particular mention of his classification of animals will be found in the chapter on Linnæus, while in what follows in this chapter attention will be confined to his observation of their structure and development and to the general influence of his work.

His great strength was in a philosophical treatment of the structure and development of animals. Professor Osborn in his interesting book, *From the Greeks to Darwin*, shows that Aristotle had thought out the essential features of evolution as a process in nature. He believed in a complete gradation from the lowest organisms to the highest, and that man is the highest point of one long and continuous ascent.

His Extensive Knowledge of Animals.—He made extensive studies of life histories. He knew that drone bees develop without previous fertilization of the eggs (by parthenogenesis); that in the squid the yolk sac of the embryo is carried in front of the mouth; that some sharks develop within the egg-tube of the mother, and in some species have a rudimentary blood-connection resembling the placenta of mammals. He had followed day by day the changes in the chick within the hen's egg, and observed the development of many other animals. In embryology also, he anticipated Harvey in appreciating the true nature of development as a process of gradual building, and not as the mere expansion of a previously formed germ. This doctrine, which is known under the name of epigenesis, was, as we shall see later, hotly contested in the eighteenth century, and has a modified application at the present time.

In reference to the structure of animals he had described the tissues, and in a rude way analyzed the organs into their component parts. It is known, furthermore, that he prepared plates of anatomical figures, but, unfortunately, these have been lost.

In estimating the contributions of ancient writers to science, it must be remembered that we have but fragments of their works to examine. It is, moreover, doubtful whether the scientific writings ascribed to Aristotle were all from his hand. The work is so uneven that Huxlev has suggested that, since the ancient philosophers taught *viva voce*, what we have of his zoölogical writings may possibly be the notes of some of his students. While this is not known to be the case, that hypothesis enables us to understand the intimate mixture of profound observation with trivial matter and obvious errors that occur in the writings ascribed to him.

Hertwig says: "It is a matter for great regret that there have been preserved only parts of his three most important zoölogical works, '*Historia animalium*,' '*De partibus*,' and '*De generatione*,' works in which zoölogy is founded as a universal science, since anatomy and embryology, physiology and classification, find equal consideration."

Some Errors.—Dissections were little practised in his day, and it must be admitted that his observations embrace many errors. He supposed the brain to be bloodless, the arteries to carry air, etc., but he has been cleared by Huxley of the mistake so often attributed to him of supposing the heart of mammals to have only three chambers. It is alto gether probable that he is credited with a larger number of errors than is justified by the facts.

He must have had unusual gifts in the exposition of these technical subjects; indeed, he made his researches appear so important to his royal patron, Alexander, that he was aided in the preparation of his great Natural History bv a grant of 800 talents (equivalent to $200,000) and by numerous assistants and collectors. Thus in ancient times was anticipated the question that is being agitated to-day—that of the support and the endowment of research.

Personal Appearance.—Some idea of his looks may be gained from Fig. 1. This is a copy of a bas-relief found in the collection of Fulvius Ursinus (d. 1600), and was originally published by J. Faber. Its authenticity as a portrait is attested (1811) by Visconti, who says that it has a perfect resemblance to the head of a small bust upon the base of which the name of Aristotle is engraved. Portrait busts and statues of Aristotle were common in ancient times. The picture of him most familiar to general readers is the copy of the head and shoulders of an ancient statue representing him with a draping over the left shoulder. This is an

FIG. 1.—ARISTOTLE, 384–322 B.C.

attractive portrait, showing a face of strong intellectuality. Its authenticity, however, is not as well established as that of the picture shown here. Other pictures, believed to be those of Aristotle, represent him later in life with receding hair, and one exists in which his baldness is very extensive. He was described as short in stature, with spindling legs and small, penetrating eyes, and to have been, in his younger days, vain and showy in his dress.

He was early left an orphan with a considerable fortune; and there are stories of early excesses after coming into his property. These charges, however, lack trustworthy support, and are usually regarded as due mainly to that undermining gossip which follows one holding prominent place and enviable recognition. His habits seem to have been those of a diligent student with a zest in his work; he was an omnivorous reader, and Plato called him the mind of his school. His large private library and his manner of living bespeak the conserving of his property, rather than its waste in selfish indulgences.

His Influence.—The influence of Aristotle was in the right direction. He made a direct appeal to nature for his facts, and founded his Natural History only on observation of the structure, physiology, and development of animals. Unfortunately, the same cannot be said of his successors.

Galen, who is mentioned above in connection with Aristotle, was a medical writer and the greatest anatomist of antiquity. On account of the relation of his work to the growth of anatomy, however, the consideration of it is reserved for the chapter on Vesalius.

Soon after the period of Aristotle the center of scientific investigation was transferred to Alexandria, where Ptolemy had erected a great museum and founded a large public library. Here mathematics and geography flourished, but natural history was little cultivated.

In order to find the next famous naturalist of antiquity, it is necessary to look to Rome. Rome, although great in political power, never became a true culture center, characterized by originality. All that remains of their thought shows us that the Roman people were not creative. In the capital of the empire, the center of its life, there arose no great scientific investigator.

Pliny.—The situation is represented by Pliny the Elder

(23–79 A.D.), the Roman general and littérateur (Fig. 2). His works on natural history, filling thirtv-seven volumes, have been preserved with greater completeness than those of other ancient writers. Their overwhelming bulk seems to have produced an impression upon those who, in the nineteenth century, heralded him as the greatest naturalist of

FIG. 2.—PLINY, 23–79 A.D.

antiquity. But an examination of his writings shows that he did nothing to deepen or broaden the knowledge of nature, and his Natural History marks a distinct retrograde movement. He was, at best, merely a compiler—"a collector of anecdotes"—who, forsaking observation, indiscriminately mixed fable, fact, and fancy taken from the writings of others. He emphasized the feature of classification which Aristotle had held in proper subordination, and he replaced the classification of Aristotle, founded on plan of organization, bv a highly artificial one, founded on the incidental circumstance of the abodes of animals—either in air, water, or on the earth.

The Arrest of Inquiry and its Effects.—Thus, natural history, transferred from a Greek to a Roman center, was already on the decline in the time of Pliny; but it was destined to sink still lower. It is an old, oft-repeated story how, with the overthrow of ancient civilization, the torch of learning was nearly extinguished. Not only was there a complete political revolution; there was also a complete change in the mental interests of mankind. The situation is so complex that it is difficult to state it with clearness. So far as science is concerned, its extinction was due to a turning away from the external world, and a complete arrest of inquiry into the phenomena of nature. This was an important part of that somber change which came over all mental life.

One of the causes that played a considerable part in the cessation of scientific investigation was the rise of the Christian church and the dominance of the priesthood in all intellectual as well as in spiritual life. The world-shunning spirit, so scrupulously cultivated by the early Christians, prompted a spirit which was hostile to observation. The behest to shun the world was acted upon too literally. The eyes were closed to nature and the mind was directed toward spiritual matters, which truly seemed of higher importance. Pres ently, the observation of nature came to be looked upon as proceeding from a prying and impious curiosity.

Books were now scarcer than during the classical period; the schools of philosophy were reduced, and the dissemination of learning ceased. The priests who had access to the books assumed direction of intellectual life. But they were largely employed with the analysis of the supernatural, without the wholesome check of observation and experiment; mystical explanations were invented for natural phenomena, while metaphysical speculation became the dominant form of mental activity.

Authority Declared the Source of Knowledge.—In this atmosphere controversies over trivial points were engendered, and the ancient writings were quoted as sustaining one side or the other. All this led to the referring of questions as to their truth or error to authority as the source of knowledge, and resulted in a complete eclipse of reason. Amusing illustrations of the situation are abundant; as when, in the Middle Ages, the question of the number of teeth in the horse was debated with great heat in many contentious writings. Apparently none of the contestants thought of the simple expedient of counting them, but tried only to sustain their position by reference to authority. Again, one who noticed spots on the sun became convinced of the error of his eyes because Aristotle had somewhere written "The face of the sun is immaculate."

This was a barren period not only for science, but also for ecclesiastical advance. Notwithstanding the fact that for more than a thousand years the only new works were written by professional theologians, there was no substantial advance in their field, and we cannot escape the reflection that the reciprocal action of free inquiry is essential to the growth of theology as of other departments of learning.

In the period from the downfall of Rome to the revival of learning, one eminent theologian, St. Augustine, stands in relief for the openness of his mind to new truth and for his expressions upon the relation of revelation in the Scriptures to the observation of nature. His position will be more clearly indicated in the chapter dealing with the rise of evolutionary thought.

Perhaps it has been the disposition of historians to paint the Middle Ages in too dark colors in order to provide a background on which fitly to portray the subsequent awakening. It was a remolding period through which it was necessary to pass after the overthrow of ancient civilization and the mixture of the less advanced people of the North with those of the South. The opportunities for advance were greatly circumscribed; the scarcity of books and the lack of facilities for travel prevented any general dissemination of learning, while the irresponsible method of the time, of appealing to authority on all questions, threw a barrier across the stream of progress. Intellectuality was not, however, entirely crushed during the prevalence of these conditions. The medieval philosophers were masters of the metaphysical method of argument, and their mentality was by no means dull. While some branches of learning might make a little advance, the study of nature suffered the most, for the knowledge of natural phenomena necessitates a mind turned outward in direct observation of the phenomena of the natural and physical universe.

Renewal of Observation.—It was an epoch of great importance, therefore, when men began again to observe, and to attempt, even in an unskilful way, hampered by intellectual inheritance and habit, to unravel the mysteries of nature and to trace the relation between causes and effects in the universe. This new movement was a revolt of the intellect against existing conditions. In it were locked up all the benefits that have accrued from the development of modern science. Just as the decline had been due to many causes, so also the general revival was complex. The invention of printing, the voyages of mariners, the rise of universities, and the circulation of ideas consequent upon the Crusades, all helped to disseminate the intellectual ferment. These generic influences aided in molding the environment, but, just as the pause in science had been due to the turning away from nature and to new mental interests, so the revival was a return to nature and to the method of science. The pioneers had to be men of determined independence; they labored against self-interest as well as opposition from the church and the priesthood, and they withstood the terrors of the Inquisition and the loss of recognition and support.

In this uncongenial atmosphere men like Galileo, Descartes, and Vesalius established the new movement and overthrew the reign of authority. With the coming of Vesalius the new era of biological progress was opened, but its growth was a slow one; a growth of which we are now to be concerned in tracing the main features.

The Epochs in Biological History

It will be helpful to outline the great epochs of biological progress before taking them up for fuller consideration. The foundation of progress was the renewal of observation in which, as already stated, all modern science was locked up.

It was an epoch in biological history when Vesalius overthrew the authority of Galen, and studied at first hand the organization of the human body.

It was an epoch when William Harvey, by adding experiment to observation, demonstrated the circulation of the blood and created a new physiology. The two coördinate branches of biology were thus early outlined.

The introduction of the microscope, mainly through the labors of Grew, Hooke, Malpighi, and Leeuwenhoek, opened a new world to the investigator, and the work of these men marks an epoch in the progress of independent inquiry.

Linnæus, by introducing short descriptions and uniform names for animals and plants, greatly advanced the subject of natural history.

Cuvier, by founding the school of comparative anatomy, so furthered the knowledge of the organization of animals that he created an epoch.

Bichat, his great contemporary, created another by laying the foundation of our knowledge of the structure of animal tissues.

Von Baer, bv his studies of the development of animal life, supplied what was lacking in the work of Cuvier and Bichat and originated modern embryology.

Haller, in the eighteenth, and Johannes Müller in the nineteenth century, so added to the ground work of Harvey that physiology was made an independent subject and was established on modern lines.

With Buffon, Erasmus Darwin, and Lamarck began an epoch in evolutionary thought which had its culminating point in the work of Charles Darwin.

After Cuvier and Bichat came the establishing of the cell-theory, which created an epoch and influenced all further progress.

Finally, through the discovery of protoplasm and the recognition that it is the seat of all vital activity, arrived the epoch which brought us to the threshold of the biology of the present day.

Step by step naturalists have been led from the obvious and superficial facts about living organisms to the deep-lying basis of all vital manifestations.

CHAPTER II

VESALIUS AND THE OVERTHROW OF AUTHORITY IN SCIENCE

Vesalius, although an anatomist, is to be recognized in a broad sense as one of the founders of biology. When one is attempting to investigate animal and plant life, not only must he become acquainted with the external appearance of living organisms, but also must acquire early a knowledge of their structure, without which other facts relating to their lives can not be disclosed. Anatomy, which is the science of the structure of organized beings, is therefore so funda mental that we find ourselves involved in tracing the history of its rise as one part of the story of biology. But it is not enough to know how animals and plants are constructed; we must also know something about the purpose of the structures and of the life that courses through them, and, accordingly, after considering the rise of anatomy, we must take a similar view of its counterpart, physiology.

The great importance of Vesalius in the history of science lies in the fact that he overthrew adherence to authority as the method of ascertaining truth, and substituted therefor observation and reason. Several of his forerunners had tried to accomplish the same end, but they had failed. He was indebted to them as everv man is indebted to his forebears, but at the same time we can not fail to see that Vesalius was worthy of the victory. He was more resolute and forceful than any of his predecessors. He was one of those rare spirits who see new truth with clearness, and have the bravery to force their thoughts on an unsympathetic public.

The Beginning of Anatomy.—In order to appreciate his service it is necessary to give a brief account of his predecessors, and of the condition of anatomy in his time. Remembering that anatomy embraces a knowledge of the architecture of all animals and plants, we can, nevertheless, see why in early times its hould have had more narrow boundaries. The medical men were the first to take an interest in the structure of the human body, because a knowledge of it is necessary for medicine and surgery. It thus happens that the earliest observations in anatomy were directed toward making known the structure of the human body and that of animals somewhat closely related to man in point of structure. Anatomical studies, therefore, began with the more complex animals instead of the simpler ones, and, later, when comparative anatomy began to be studied, this led to many misunderstandings; since the structure of man became the type to which all others were referred, while, on account of his derivation, his structure presents the greatest modification of the vertebrate type.

It was so difficult in the early days to get an opportunity to study the human body that the pioneer anatomists were obliged to gain their knowledge by dissections of animals, as the dog, and occasionally the monkey. In this way Aristotle and his forerunners learned much about anatomy. About 300 B.C., the dissection of the human body was legalized in the Alexandrian school, the bodies of condemned criminals being devoted to that purpose. But this did not become general even for medical practitioners, and anatomy continued to be studied mainly from brute animals.

Galen.—The anatomist of antiquity who outshines all others was Galen (Claudius Galenus, 130–200 A.D.), who lived some time in Pergamos, and for five years in Rome, during the second century of the Christian era. He was a man of much talent, both as an observer and as a writer. His descriptions were clear and forceful, and for twelve centuries his works exerted the greatest influence of those of all scientific writers. In his writings was gathered all the anatomical knowledge of his predecessors, to which he had added observations of his own. He was a man of originality, but not having the human body for dissection, he erred in expounding its structure "on the faith of observations made on lower animals." He used the right method in arriving at his facts. Huxley says: "No one can read Galen's works without being impressed with the marvelous extent and diversity of his knowledge, and by his clear grasp of those experimental methods by which alone physiology can be advanced."

Anatomy in the Middle Ages.—But now we shall see how the arrest of inquiry already spoken of operated in the field of anatomy. The condition of anatomv in the Middle Ages was the condition of all science in the same period. From its practical importance anatomy had to be taught to medical men, while physics and chemistry, biology and comparative anatomy remained in an undeveloped state. The way in which this science was taught is a feature which characterizes the intellectual life of the Middle Ages. Instead of having anatomy taught by observations, the writings of Galen were expounded from the desk, frequently without demonstrations of any kind. Thus his work came to be set up as the one unfailing authority on anatomical knowledge. This was in accord with the dominant ecclesiastical influence of the time. Reference to authority was the method of the theologians, and by analogy it became the method of all learning. As the Scriptures were accepted as the unfailing guide to spiritual truth, so Galen and other ancient writers were made the guides to scientific truth and thought. The baneful effects of this in stifling inquiry and in reducing knowledge

FIG. 3.—GALEN, 131–200.

From *Acta Medicorum Berolinensium*, 1715.

to parrot-like repetition of ancient formulas are so obvious that they need not be especially dwelt upon.

Predecessors of Vesalius.—Italy gave birth to the first anatomists who led a revolt against this slavery to authority in scientific matters. Of the eminent anatomists who preceded Vesalius it will be necessary to mention only three. Mundinus, or Mondino, professor at the University of Bologna, who, in the early part of the fourteenth century, dissected three bodies, published in 1315 a work founded upon human dissection. He was a man of originality whose work created a sensation in the medical world, but did not supersede Galen's. His influence, although exerted in the right direction, was not successful in establishing observation as the method of teaching anatomy. His book, however, was sometimes used as an introduction to Galen's writings or in conjunction with them.

The next man who requires notice is Berengarius of Carpi, who was a professor in the University of Bologna in the early part of the sixteenth century. He is said to have dissected not less than one hundred human bodies; and although his opportunities for practical study were greater than those of Mondino, his attempts to place the science of anatomy upon a higher level were also unsuccessful.

We pass now from Italy to France, where Sylvius (1478–1555), one of the teachers of Vesalius, made his mark. His name is preserved to-day in the *fissure of Sylvius* in the brain, but he was not an original investigator, and he succeeded only in "making a reputation to which his researches do not entitle him." He was a selfish, avaricious man whose adoption of anatomy was not due to scientific interest, but to a love of gain. At the age of fifty he forsook the teaching of the classics for the money to be made by teaching anatomy. He was a blind admirer of Galen, and read his works to medical students without dissections, except that from time to time dogs were brought into the amphitheater and their structure exposed by unskilled barbers.

Vesalius.—Vesalius now came upon the scene; and through his efforts, before he was thirty years of age, the idol of authority had been shattered, and, mainly through his persistence, the method of so great moment to future ages had been established. He was well fitted to do battle against tradition—strong in body, in mind, and in purpose, gifted and forceful; and, furthermore, his work was marked by concentration and by the high moral quality of fidelity to truth.

Vesalius was born in Brussels on the last day of the year 1514, of an ancestry of physicians and learned men, from whom he inherited his leaning toward scientific pursuits. Early in life he exhibited a passion for anatomy; he dissected birds, rabbits, dogs, and other animals. Although having a strong bent in this direction, he was not a man of single talent. He was schooled in all the learning of his time, and his earliest publication was a translation from the Greek of the ninth book of *Rhazes*. After his early training at Brussels and at the University of Louvain, in 1533, at the age of 18, he went to Paris to study medicine, where, in anatomy, he came under Sylvius and Günther.

His Force and Independence.—His impetuous nature was shown in the amphitheatre of Sylvius, where, at the third lecture, he pushed aside the clumsy surgeon barbers, and himself exposed the parts as they should be. He could not be satisfied with the exposition of the printed page; he must see with his own eyes, must grasp through his own experience the facts of anatomical structure. This demand of his nature shows not only how impatient he was with sham, but also how much more he was in touch with reality than were the men of his time.

After three years at the French capital, owing to wars

in Belgium, he went back to Louvain without obtaining his medical degree. After a short experience as surgeon on the field of battle, he went to Padua, whither he was attracted by reports of the opportunities for practical dissection that he so much desired to undertake. There his talents were recognized, and just after receiving his degree of Doctor of Medicine in 1537, he was given a post in surgery, with the care of anatomy, in the university.

His Reform of the Teaching of Anatomy.—The sympathetic and graphic description of this period of his career by Sir Michael Foster is so good that I can not refrain from quoting it: "He at once began to teach anatomy in his own new way. Not to unskilled, ignorant barbers would he entrust the task of laying bare before the students the secrets of the human frame; his own hand, and his own hand alone, was cunning enough to track out the pattern of the structures which day by day were becoming more clear to him. Following venerated customs, he began his academic labors by 'reading' Galen, as others had done before him, using his dissections to illustrate what Galen had said. But, time after time, the body on the table said something different from that which Galen had written.

"He tried to do what others had done before him—he tried to believe Galen rather than his own eyes, but his eyes were too strong for him; and in the end he cast Galen and his writings to the winds, and taught only what he himself had seen and what he could make his students see, too. Thus he brought into anatomy the new spirit of the time, and the men of the time, the young men of the time, answered the new voice. Students flocked to his lectures; his hearers amounted, it is said, to some five hundred, and an enlightened senate recognized his worth by repeatedly raising his emoluments.

"Five years he thus spent in untiring labors at Padua.

FIG. 4.—VESALIUS, 1514–1564.

Five years he wrought, not weaving a web of fancied thought, but patiently disentangling the pattern of the texture of the human body, trusting to the words of no master, admitting nothing but that which he himself had seen; and at the end of the five years, in 1542, while he was as yet not twenty-eight years of age, he was able to write the dedication to Charles V of a folio work entitled the 'Structure of the Human Body,' adorned with many plates and woodcuts which appeared at Basel in the following year, 1543."

His Physiognomy.—This classic with the Latin title, *De Humani Corporis Fabrica*, requires some special notice· but first let us have a portrait of Vesalius, the master. Fig. 4 shows a reproduction of the portrait with which his work is provided. He is represented in academic costume, probably that which he wore at lectures, in the act of demonstrating the muscles of the arm. The picture is reduced, and in the reduction loses something of the force of the original. We see a strong, independent, self-willed countenance; what his features lack in refinement they make up in force; not an artistic or poetic face, but the face of the man of action with scholarly training.

His Great Book.—The book of Vesalius laid the foundation of modern biological science. It is more than a landmark in the progress of science—it created an epoch. It is not only interesting historically, but on account of the highly artistic plates with which it is illustrated it is interesting to examine by one not an anatomist. For executing the plates Vesalius secured the service of a fellow-countryman, John Stephen de Calcar, who was one of the most gifted pupils of Titian. The drawings are of such high artistic quality that for a long time they were ascribed to Titian. The artist has attempted to soften the necessarily prosaic nature of anatomical illustrations by introducing an artistic background of landscape of varied features, with bridges, roads, streams,

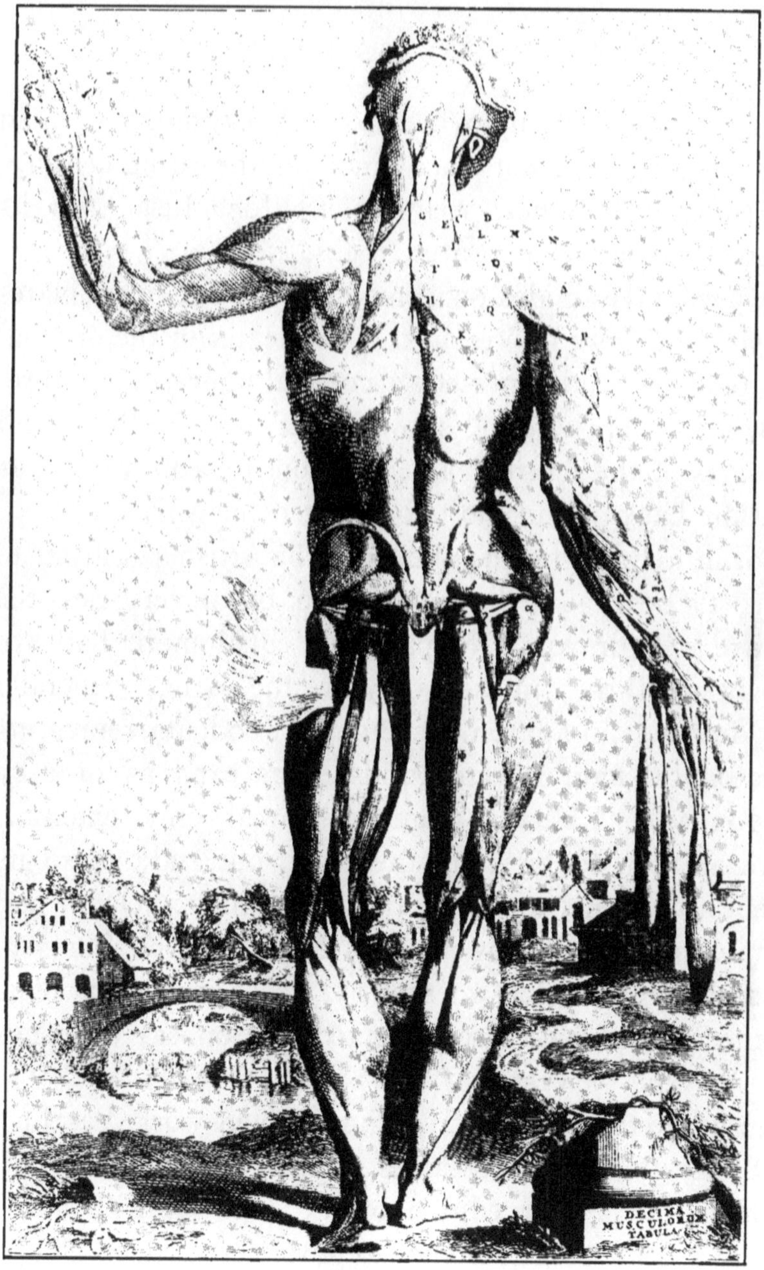

FIG. 5.—Anatomical Sketch from Vesalius's *Fabrica*.
(Photographed and reduced from the facsimile edition of 1728.)

buildings, etc. The employment of a background even in portrait-painting was not uncommon in the same century, as in Leonardo da Vinci's well-known Mona Lisa, with its suggestive perspective of water, rocks, etc.

Fig. 5 will give an idea on a small scale of one of the plates illustrating the work of Vesalius. The plates in the original are of folio size, and represent a colossal figure in the foreground, with a background showing between the limbs and at the sides of the figure. There is considerable variety as regards the background, no two plates being alike.

Also, in delineating the skeleton, the artist has given to it an artistic pose, as is shown in Fig. 6, but nevertheless the bones are well drawn. No plates of equal merit had appeared before these; in fact, they are the earliest generally known drawings in anatomy, although woodcuts representing anatomical figures were published as early as 1491 by John Ketham. Ketham's figures showed only externals and preparations for opening the body, but rude woodcuts representing internal anatomy and the human skeleton had been published notably by Magnus Hundt, 1501; Phrysen, 1518; and Berengarius, 1521 and 1523. Leonardo da Vinci and other artists had also executed anatomical drawings before the time of Vesalius.

Previous to the publication of the complete work, Vesalius, in 1538, had published six tables of anatomy, and, in 1555, he brought out a new edition of the *Fabrica*, with slight additions, especially in reference to physiology, which will be adverted to in the chapter on Harvey.

In the original edition of 1543 the illustrations are not collected in the form of plates, but are distributed through the text, the larger ones making full-page (folio) illustrations. In this edition also the chapters are introduced with an initial letter showing curious anatomical figures in miniature, some of which are shown in Fig. 7.

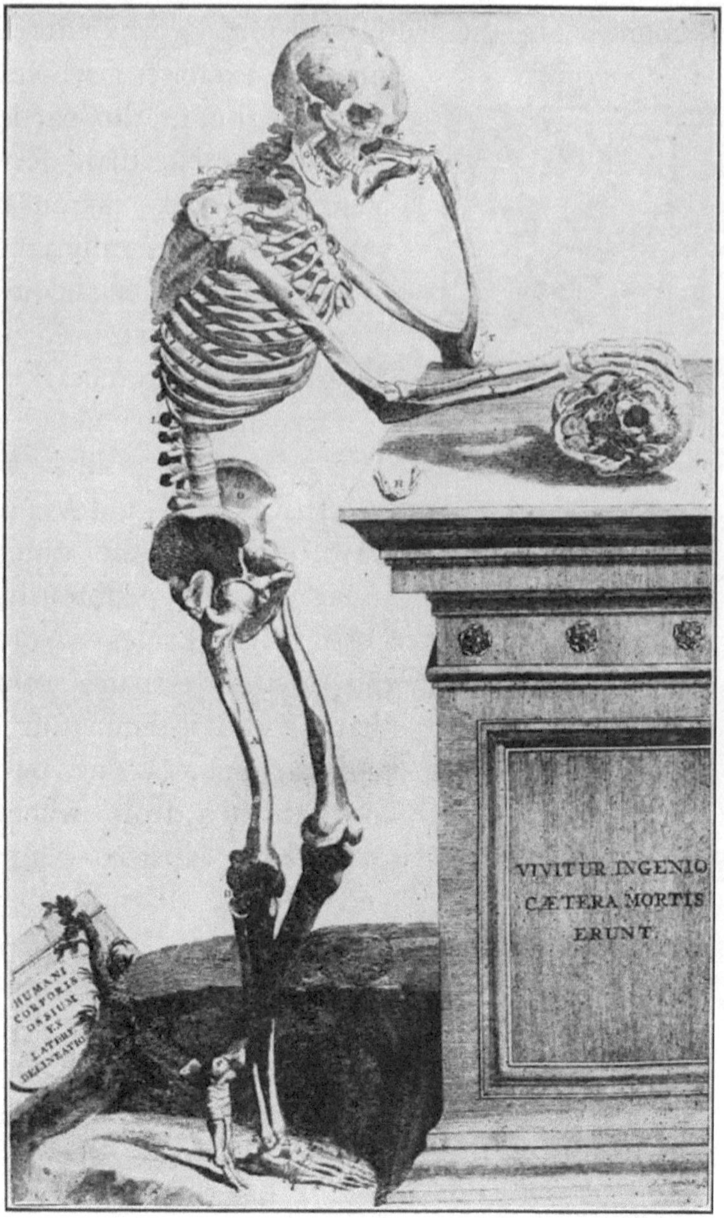

FIG. 6.—The Skeleton, from Vesalius's *Fabrica*.

The *Fabrica* of Vesalius was a piece of careful, honest work, the moral influence of which must not be overlooked. At any moment in the world's history, work marked by sinceritv exercises a wholesome influence, but at this particular stage of intellectual development such work was an innovation, and its significance for progress was wider and deeper than it might have been under different circumstances.

FIG. 7.—Initial letters from Vesalius's *Fabrica* of 1543.

Opposition to Vesalius. — The beneficent results of his efforts were to unfold afterward, since, at the time, his utterances were vigorously opposed from all sides. Not only did the ecclesiastics contend that he was disseminating false and harmful doctrine, but the medical men from whom he might have expected svmpathy and support violently opposed his teachings.

Many amusing arguments were brought forward to discredit Vesalius, and to uphold the authority of Galen. Vesalius showed that in the human body the lower jaw is a single bone—that it is not divided as it is in the dog and other lower mammals, and, as Galen had taught, also in the human subjects. He showed that the sternum, or breast bone, has three parts instead of eight; he showed that the thigh bones are straight and not curved, as they are in the dog. Sylvius, his old teacher, was one of his bitterest opponents; he declared that the human body had undergone changes in structure since the time of Galen, and, with the object of defending the ancient anatomist, "he asserted that the straight thigh bones, which, as everv one saw, were not curved in accordance with the teaching of Galen, were the result of the narrow trousers of his contemporaries, and that thev must have been curved in their natural condition, when uninterfered with by art!"

The theologians also found other points for contention. It was a widely accepted dogma that man should have one less rib on one side, because from the Scriptural account Eve was formed from one of Adam's ribs. This, of course, Vesalius did not find to be the case. It was also generally believed at this time that there was in the body an indestructible resurrection-bone which formed the nucleus of the resurrection-body. Vesalius said that he would leave the question of the existence of such a bone to be decided by the theologians, as it did not appear to him to be an anatomical question.

The Court Physician.—The hand of the church was heavy upon him, and the hatred shown in attacks from various quarters threw Vesalius into a state of despondency and anger. In this frame of mind he destroyed manuscripts upon which he had expended much labor. His disappointment in the reception of his work probably had much to do in deciding him to relinquish his professorship and accept the post of court physician to Charles V of the United Kingdoms of Spain and Belgium. After the death of Charles, he remained with Philip II, who succeeded to the throne. Here he waxed rich and famous, but he was always under suspicion by the clerical powers, who from time to time found means of discrediting him. The circumstances of his leaving Spain are not definitely known. One account has it that he made a *post-mortem* examination of a body which showed signs of life during the operation, and that he was required to undertake a pilgrimage to the Holy Land to clear his soul of sacrilege. Whether or not this was the reason is uncertain, but after nineteen years at the Spanish Court he left, in 1563 and journeyed to Jerusalem. On his return from Palestine he suffered shipwreck and died from the effects of exposure on Zanti, one of the Ionian Islands. It is also said that while on this pilgrimage he had been offered the position of professor of anatomy as successor to Fallopius, who had died in 1563, and that, had he lived, he would have come back honorablv to his old post.

Eustachius and Fallopius.—The work of two of his contemporaries, Eustachius and Fallopius, requires notice. Cuvier says in his *Histoire des Sciences Naturelles* that those three men were the founders of modern anatomy. Vesalius was a greater man than either of the other two, and his influence was more far-reaching. He reformed the entire field of anatomy, while the names of Eustachius and Fallopius are connected especially with a smaller part of the field. Eustachius described the Eustachian tube of the ear and gave especial attention to sense organs; Fallopius made special investigations upon the viscera, and described the Fallopian tube.

Fallopius was a suave, polite man, who became professor of anatomy at Padua; he opposed Vesalius, but his attacks were couched in respectful terms.

Eustachius, the professor of anatomy at Rome, was of a different type, a harsh, violent man, who assailed Vesalius with virulence. He corrected some mistakes of Vesalius, and prepared new plates on anatomy, which, however, were not published until 1754, and therefore did not exert the influence upon anatomical studies that those of Vesalius did.

The Especial Service of Vesalius.—It should be remembered that both these men had the advantage of the sketches made under the direction of Vesalius. Pioneers and pathbreakers are under special limitations of being in a new territorv, and make more errors than they would in following

FIG. 8.—FALLOPIUS, 1523–1563.

another's survey of the same territory; it takes much less creative force to correct the errors of a first survey than to make the original discoveries. Everything considered, Vesalius is deserving of the position assigned to him. He was great in a larger sense, and it was his researches in particular which re-established scientific method and made further progress possible. His errors were corrected, not by an appeal to authority, but by the method which he founded. His great claim to renown is, not that his work outshone all other work (that of Galen in particular) in accuracy and brilliancy, but that he overthrew dependence on authority and re-established the scientific method of ascertaining truth. It was the method of Aristotle and Galen given anew to the world.

The spirit of progress was now released from bondage, but we have still a long way to go under its guidance to reach the gateway of modern biology.

CHAPTER III

WILLIAM HARVEY AND EXPERIMENTAL OBSERVATION

AFTER the splendid observations of Vesalius, revealing in a new light the construction of the human body, Harvey took the next general step by introducing experiment to determine the use or purpose of the structures that Vesalius had so clearly exposed. Thus the work of Harvey was complemental to that of Vesalius, and we may safely say that, taken together, the work of these two men laid the foundations of the modern method of investigating nature. The results they obtained, and the influence of their method, are of especial interest to us in the present connection, inasmuch as they stand at the beginning of biological science after the Renaissance. Although the observations of both were applied mainly to the human body, they served to open the entire field of structural studies and of experimental observations on living organisms.

Many of the experiments of Harvey, notablv those relating to the movements of the heart, were, of course, conducted upon the lower animals, as the frog, the dog, etc. His experiments on the living human body consisted mainly in applying ligatures to the arms and the legs. Nevertheless, the results of all his experiments related to the phenomena of the circulation in the human body, and were primarily for the use of medical men.

In what sense the observations of the two men were complemental will be better understood when we remember that there are two aspects in which living organisms should always be considered in biological studies; first, the structure, and, then, the use that the structures subserve. One view is essential to the other, and no investigation of animals and plants is complete in which the two ideas are not involved. Just as a knowledge of the construction of a machine is necessary to understand its action, so the anatomical analysis of an organ must precede a knowledge of its office. The term "physiological anatomy of an organ," so commonlv used in text-books on physiology, illustrates the point. We can not appreciate the work of such an organ as the liver without a knowledge of the arrangement of its working units. The work of the anatomist concerns the statics of the body, that of the physiologist the dynamics; properly combined, they give a complete picture of the living organism.

It is to be remembered that the observations of Vesalius were not confined exclusivelv to structure; he made some experiments and some comments on the use of parts of the body, but his work was mainly structural, while that which distinguishes Harvey's research is inductions founded on experimental observation of the action of living tissues.

The service of Vesalius and Harvey in opening the path to biological advance is very conspicuous, but thev were not the only pioneers; their work was a part of the general revival of science in which Galileo, Descartes, and others had their part. While the birth of the experimental method was not due to the exertions of Harvey alone, nevertheless it should stand to his credit that he established that method in biological lines. Aristotle and Galen both had employed experiments in their researches, and Harvey's step was in the nature of a revival of the method of the old Greeks.

Harvey's Education.—Harvey was fitted both by native talent and by his training for the part which he played in the intellectual awakening. He was born at Folkestone, on the south coast of England, in 1578, the son of a prosperous yeoman. The Harvey familv was well esteemed, and the father of William was at one time the mayor of Folkestone. Young Harvey, after five years in the King's school at Canterbury, went to Cambridge, and in 1593, at the age of sixteen, entered Caius College. He had already shown a fondness for observations upon the organization of animals, but it is unlikely that he was able to cultivate this at the university. There his studies consisted mainly of Latin and Greek, with some training in debate and elementary instruction in the science of physics.

At Padua.—In 1597, at the age of nineteen, he was graduated with the Arts degree, and the following year he turned his steps toward Italy in search of the best medical instruction that could be found at that time in all the world. He selected the great university of Padua as his place of sojourn, being attracted thither by the fame of some of its medical teachers. He was particularly fortunate in receiving his instruction in anatomy and physiology from Fabricius, one of the most learned and highly honored teachers in Italy. The fame of this master of medicine, who, from his birthplace, is usually given the full name of Fabricius *ab Aquapendente*, had spread to the intellectual centers of the world, where his work as anatomist and surgeon was especially recognized. A fast friendship sprang up between the young medical student and this ripe anatomist, the influence of which must have been very great in shaping the future work of Harvey.

Fabricius was already sixty-one years of age, and when Harvey came to Padua was perfecting his knowledge upon the valves of the veins. The young student was taken fully into his confidence, and here was laid that first familiarity with the circulatory system, the knowledge of which Harvey was destined so much to advance and amplify. But it was the stimulus of his master's friendship, rather than what he taught about the circulation, that was of assistance to Harvey. For the views of Fabricius in reference to the circulation were those of Galen; and his conception of the use of the valves of the veins was entirely wrong. A portrait of this great teacher of Harvev is shown in Fig. 9.

At Padua young Harvey attracted notice as a student of originality and force, and seems to have been a favorite with the student body as well as with his teachers. His position in the university may be inferred from the fact that he belonged to one of the aristocratic-student organizations, and, further, that he was designated a "councilor" for England. The practice of having student councilors was then in vogue in Padua; the students comprising the council met for deliberations, and very largely managed the university by their votes upon instructors and university measures.

It is a favorable comment upon the professional education of his time that, after graduating at the University of Cambridge, he studied four or more years (Willis says five years) in scientific and medical lines to reach the degree of Doctor of Physic.

On leaving Padua, in 1602, he returned to England and took the examinations for the degree of M.D. from Cambridge, inasmuch as the medical degree from an English university advanced his prospects of receiving a position at home. He opened practice, was married in 1604, and the same year began to give public lectures on anatomy.

His Personal Qualities.—Harvey had marked individuality, and seems to have produced a powerful impression upon those with whom he came in contact as one possessing unusual intellectual powers and independence of character. He inspired confidence in people, and it is significant that, in reference to the circulation of the blood, he won to his way of thinking his associates in the medical profession. This is important testimony as to his personal force, since his ideas were opposed to the belief of the time, and since also away from home they were vigorously assailed.

FIG. 9.—FABRICIUS, 1537–1619, HARVEY'S TEACHER.

Although described as choleric and hasty, he had also winning qualities, so that he retained warm friendships throughout his life, and was at all times held in high respect.

FIG. 10.—WILLIAM HARVEY, 1578–1667.

It must be said also that in his replies to his critics, he showed great moderation.

The contemplative face of Harvey is shown in Fig. 10. This is taken from his picture in the National Portrait Gallery in London, and is usually regarded as the second-best portrait of Harvey, since the one painted by Jansen, now in possession of the Royal College of Physicians, is believed to be the best one extant. The picture reproduced here shows a countenance of composed intellectual strength, with a suggestion, in the forehead and outline of the face, of some of the portraits of Shakespeare.

An idea of his personal appearance may be had from the description of Aubrey, who says: "Harvey was not tall, but of the lowest stature; round faced, with a complexion like the wainscot; his eyes small, round, very black, and full of spirit; his hair black as a raven, but quite white twenty years before he died; rapid in his utterance, choleric, given to gesture," etc.

He was less impetuous than Vesalius, who had published his work at twenty-eight; Harvey had demonstrated his ideas of the circulation in public anatomies and lectures for twelve years before publishing them, and when his great classic on the Movement of the Heart and Blood first appeared in 1628, he was alreadv fifty years of age. This is a good example for young investigators of to-day who, in order to secure priority of announcement, so frequently rush into print with imperfect observations as preliminary communications.

Harvey's Writings.—Harvey's publications were all great; in embryology, as in physiology, he produced a memorable treatise. But his publications do not fully represent his activity as an investigator; it is known that through the fortunes of war, while connected with the sovereign Charles I as court physician, he lost manuscripts and drawings upon the comparative anatomy and development of insects and other animals. His position in embrvology will be dealt with in the chapter on the Development of Animals, and he will come up for consideration again in the chapter on the Rise of Physiology. Here we are concerned chiefly with his general influence on the development of biology.

His Great Classic on Movement of the Heart and Blood.—Since his book on the circulation of the blood is regarded as one of the greatest monuments along the highroad of biology, it is time to make mention of it in particular. Although relatively small, it has a long title out of proportion to its size: *Exercitatio Anatomica de Motu Cordis et Sanguinis in Animalibus*, which may be freely translated, "An Anatomical Disquisition on the Movement of the Heart and Blood in Animals." The book is usually spoken of under the shorter title, *De Motu Cordis et Sanguinis.* The full title seems somewhat repellent, but the contents of the book will prove to be interesting to general readers. It is a clear, logical demonstration of the subject, proceeding with directness from one point to another until the culminating force of the argument grows complete and convincing.

The book in its first edition was a quarto volume of seventy-eight pages, published in Frankfort in 1628. An interesting facsimile reprint of this work, translated into English, was privately reproduced in 1894 by Dr. Moreton and published in Canterbury. As stated above, it is known that Harvey had presented and demonstrated his views in his lectures since 1616. In his book he showed for the first time ever in print, that all the blood in the body moves in a circuit, and that the beating of the heart supplies the propelling force. Both ideas were new, and in order to appreciate in what sense they were original with Harvey, we must inquire into the views of his forerunners.

Question as to Harvey's Originality.—The question of how near some of his predecessors came to anticipating his demonstration of the circulation has been much debated. It has been often maintained that Servetus and Realdus Columbus held the conception of the circulation for which Harvey has become so celebrated. Of the various accounts of the views of Harvey's predecessors, those of Willis, Huxley, and Michael Foster are among the most judicial; that of Foster, indeed, inasmuch as it contains ample quotations from the original sources, is the most nearly complete and satisfactory. The discussion is too long to enter into fully here, but a brief outline is necessary to understand what he accomplished, and to put his discovery in the proper light.

To say that he first discovered—or, more properly, demonstrated—the circulation of the blood carries the impression that he knew of the existence of capillaries connecting the arteries and the veins, and had ocular proof of the circulation through these connecting vessels. But he did not actuallv see the blood moving from veins to arteries, and he knew not of the capillaries. He understood clearly from his observations and experiments that all the blood passes from veins to arteries and moves in "a kind of circle"; still, he thought that it filters through the tissues in getting from one kind of vessel to the other. It was reserved for Malpighi, in 1661, and Leeuwenhoek, in 1669, to see, with the aid of lenses, the movement of the blood through the capillaries in the transparent parts of animal tissues. (See under Leeuwenhoek, p. 84.)

The demonstration by Harvey of the movement of the blood in a circuit was a matter of cogent reasoning, based on experiments with ligatures, on the exposure of the heart in animals and the analysis of its movements. It has been commonly maintained (as by Whewell) that he deduced the circulation from observations of the valves in the veins, but this is not at all the case. The central point of Harvey's reasoning is that the quantity of blood which leaves the left cavity of the heart in a given space of time makes necessary its return to the heart, since in a half-hour (or less) the heart, by successive pulsations, throws into the great arterv more than the total quantity of blood in the body. Huxley points out that this is the first time that quantitative determinations were introduced into physiology.

Views of His Predecessors on the Movement of the Blood.—Galen's view of the movement of the blood was not completely replaced until the establishment of Harvey's view. The Greek anatomist thought that there was an ebb and flow of blood within both veins and arteries throughout the system. The left side of the heart was supposed to contain blood vitalized by a mixture of animal spirits within the lungs. The veins were thought to contain crude blood. He supposed, further, that there was a communication between the right and the left side of the heart through very minute pores in the septum, and that some blood from the right side passed through the pores into the left side and there became charged with animal spirits. It should also be pointed out that Galen believed in the transference of some blood through the lungs from the right to the left side of the heart, and in this foreshadowed the views which were later developed bv Servetus and Realdus Columbus.

Vesalius, in the first edition of his work (1543) expressed doubts upon the existence of pores in the partition-wall of the heart through which blood could pass; and in the second edition (1555) of the *Fabrica* he became more skeptical. In taking this position he attacked a fundamental part of the belief of Galen. The careful structural studies of Vesalius must have led him very near to an understanding of the connection between arteries and veins. Fig. 11 shows one of his sketches of the arrangement of arteries and veins. He saw that the minute terminals of arteries and veins came very close together in the tissues of the body, but he did not grasp the meaning of the observation, because his physiology was still that of Galen; Vesalius continued to believe that the arteries contained blood mixed with spirits, and the veins crude blood, and his idea of the movement was that of an ebb and flow. In reference to the anatomy of the blood-vessels, he goes so far as to say of the portal vein and the vena cava in the liver that "the extreme ramifications of these veins inosculate with each other, and in many places appear

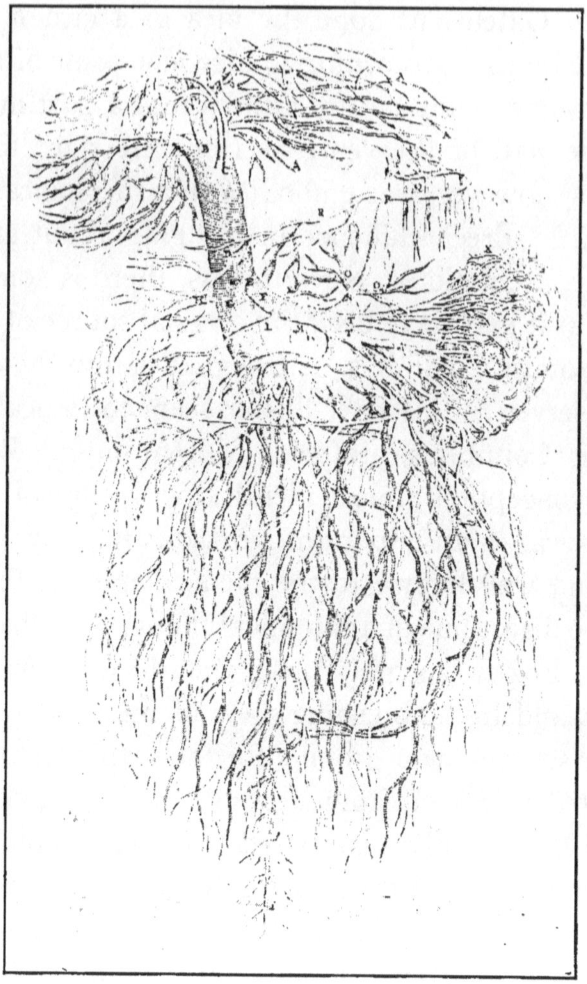

FIG. 11.—Scheme of the Portal Circulation According to Vesalius, 1543.

to unite and be continuous" All who followed him had the advantage of his drawings showing the parallel arrangement of arteries and veins, and their close approach to each other in their minute terminal twigs, but no one before Harvey fully grasped the idea of the movement of the blood in a complete circuit.

Servetus, in his work on the Restoration of Christianity (*Restitutio Christianismi*, 1553), the work for which Calvin accomplished his burning at the stake, expressed more clearly than Galen had done the idea of a circuit of blood through the lungs. According to his view, some of the blood took this course, while he still admits that a part may exude through the wall of the ventricle from the right to the left side. This, however, was embodied in a theological treatise, and had little direct influence in bringing about an altered view of the circulation. Nevertheless, there is some reason to think that it may have been the original source of the ideas of the anatomist Columbus, as the studies into the character of that observer by Michael Foster seem to indicate.

Realdus Columbus, professor of anatomy at Rome, expressed a conception almost identical with that of Servetus, and as this was in an important work on anatomy, published in 1559, and well known to the medical men of the period, it lay in the direct line of anatomical thought and had greater influence. Foster suggests that the devious methods of Columbus, and his unblushing theft of intellectual property from other sources, give ground for the suspicion that he had appropriated this idea from Servetus without acknowledgment. Although Calvin supposed that the complete edition of a thousand copies of the work of Servetus had been burned with its author in 1553, a few copies escaped, and possibly one of these had been examined by Columbus. This assumption is strengthened by the circumstance that Columbus gives no record of observations, but almost exactly repeats the words of Servetus.

Cæsalpinus, the botanist and medical man, expressed in 1571 and 1593 similar ideas of the movement of the blood (probably as a matter of argument, since there is no record of either observations or experiments by him). He also laid hold of a still more important conception, *viz.*, that some of the blood passes from the left side of the heart through the arteries of the body, and returns to the right side of the heart by the veins. But a fair consideration of the claims of these men as forerunners of Harvey requires quotations from their works and a critical examination of the evidence thus adduced. This has been excellently done by Michael Foster in his *Lectures on the History of Physiology.* Further considerations of this aspect of the question would lie beyond the purposes of this book.

At most, before Harvey, the circuit through the lungs had been vaguely defined by Galen, Servetus, Columbus, and Cæsalpinus, and the latter had supposed some blood to pass from the heart by the arteries and to return to it by the veins; but no one had arrived at an idea of a complete circulation of all the blood through the system, and no one had grasped the consequences involved in such a conception. Harvey's idea of the movement of the heart (*De Motu Cordis*) was new; his notion of the circulation (*et Sanguinis*) was new; and his method of demonstrating these was new.

Harvey's Argument.—The gist of Harvey's arguments is indicated in the following propositions quoted with slight modifications from Hall's *Physiology:* (I) The heart passively dilates and actively contracts; (II) the auricles contract before the ventricles do; (III) the contraction of the auricles forces the blood into the ventricles; (IV) the arteries have no "pulsific power," *i.e.*, they dilate passively, since the pulsation of the arteries is nothing else than the impulse of the blood within them; (V) the heart is the organ of propulsion of the blood; (VI) in passing from the right ventricle to the left auricle the blood transudes through the parenchyma of the lungs; (VII) the quantity and rate of passage of the blood peripherally from the heart makes it a physical necessity that most of the blood return to the heart; (VIII) the blood does return to the heart by way of the veins. It will be noticed that the proposition VII is the important one; in it is involved the idea of applying measurement to a physiological process.

Harvey's Influence.—Harvey was a versatile student. He was a comparative anatomist as well as a physiologist and embryologist; he had investigated the anatomy of about sixtv animals and the embryology of insects as well as of vertebrates, and his general influence in promoting biological work was extensive.

His work on the movement of the blood was more than a record of a series of careful investigations; it was a land-mark in progress. When we reflect on the part played in the body by the blood, we readily see that a correct idea of how it carries nourishment to the tissues, and how it brings away from them the products of disintegrated protoplasm is of prime importance in physiology. It is the point from which spring all other ideas of the action of tissues, and until this was known the fine analysis of vital processes could not be made. The true idea of respiration, of the secretion by glands, the chemical changes in the tissues, in fact, of all the general activities of the body, hinge upon this conception. It was these consequences of his demonstration, rather than the fact that the blood moves in a circuit, which made it so important. This discovery created modern physiology, and as that branch of inquiry is one of the parts of general biology, the bearing of Harvev's discoverv upon biological thought can be readily surmised.

Those who wish to examine Harvey's views at first hand, without the burden of translating them from the Latin, will find an edition of his complete works translated into English by Willis, and published by the Ray Society, of London.

As is always the case with new truths, there was hostility

to accepting his views. In England this hostility was slight on account of his great personal influence, but on the Continent there was many a sharp criticism passed upon his work. His views were so illuminating that they were certain of triumph, and even in his lifetime were generally accepted. Thus the new conception of vital activities, together with his method of inquiry, became permanent parts of biological science.

CHAPTER IV

THE INTRODUCTION OF THE MICROSCOPE AND THE PROGRESS OF INDEPENDENT OBSERVATION

THE introduction of the microscope greatly increased the ocular powers of observers, and, in the seventeenth century, led to many new departures. By its use the observations were carried from the plane of gross anatomy to that of minute structure; the anatomy of small forms of life, like insects, began to be studied, and also the smaller microscopic animalcula were for the first time made known.

Putting aside the disputed questions as to the time of the invention and the identity of the inventor of the microscope—whether to Fontana, Galileo, or the Jenssens belongs the credit—we know that it was improved bv the Hollander Drebbel in the early years of the seventeenth century, but was not seriously applied to anatomical studies till after the middle of that century.

THE PIONEER MICROSCOPISTS

The names especially associated with early microscopic observations are those of Hooke and Grew in England, Malpighi in Italy, and Swammerdam and Leeuwenhoek, both in Holland. Their microscopes were imperfect, and were of two kinds: simple lenses, and lenses in combination, forming what we now know as the compound microscope. Some forms of these early microscopes will be described and illustrated later. Although thus early introduced, microscopic observation did not produce its great results until the nineteenth century, just after magnifying-lenses had been greatly improved.

Robert Hooke (1635–1703), of London, published in 1665 a book of observations with the microscope entitled *Micrographia*, which was embellished with eighty-three plates of figures. Hooke was a man of fine mental endowment, who had received a good scientific training at the University of Cambridge, but who lacked fixedness of purpose in the employment of his talents. He did good work in mathematics, made many models for experimenting with flying machines, and claimed to have discovered gravitation before

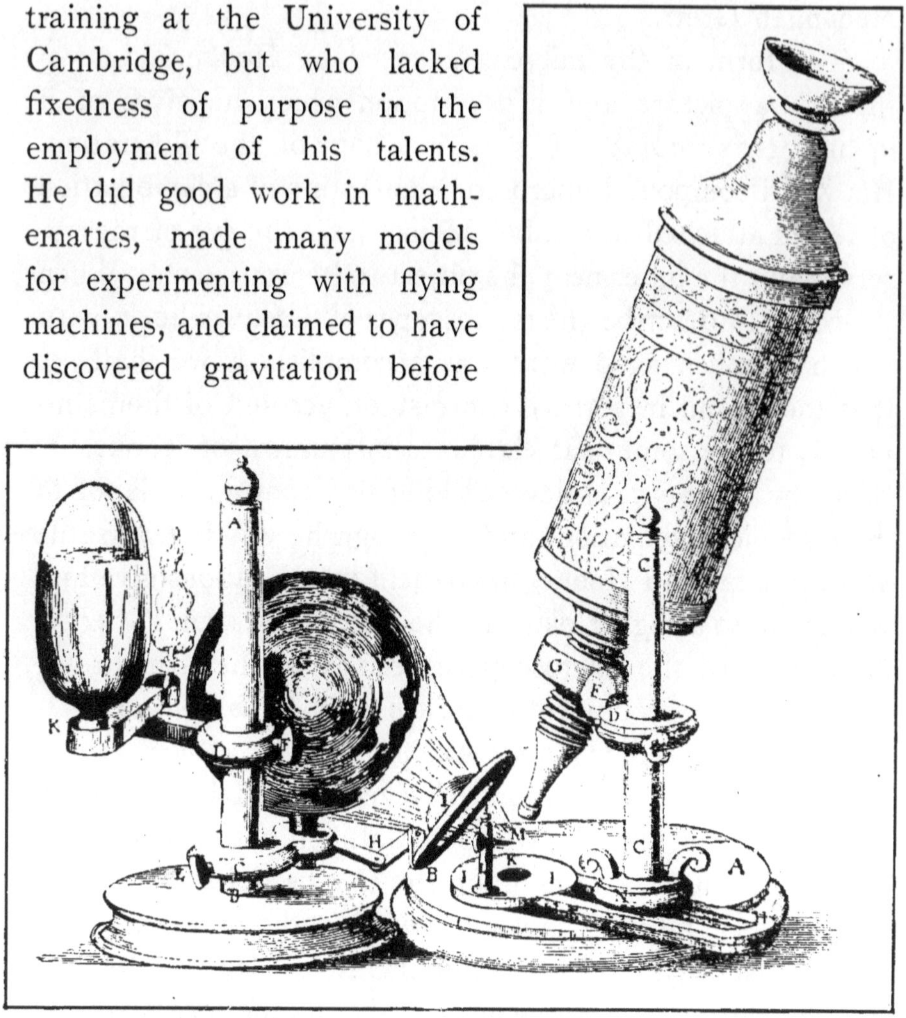

FIG. 12.—HOOKE'S MICROSCOPE, 1665.
From Carpenter's *The Microscope and Its Revelations.* Permission of P. Blakiston's Sons & Co.

Newton, and also the use of a spring for regulating watches before Huygens, etc. He gave his attention to microscopic study for a time and then dropped it; yet, although we can not accord to him a prominent place in the history of biology, he must receive mention as a pioneer worker with the microscope. His book gave a powerful stimulus to microscopy in England, and, partly through its influence, labor in this field was carried on more systematically by his fellow-countryman Nehemiah Grew.

The form of the microscope used by Hooke is known through a picture and a description which he gives of it in his *Micrographia.* Fig. 12 is a copy of the illustration. His was a compound microscope consisting of a combination of lenses attached to a tube, one set near the eve of the observer and the other near the object to be examined. When we come to describe the microscopes of Leeuwenhoek, with which so much good work was accomplished, we shall see that they stand in marked contrast, on account of their simplicity, to the somewhat elaborate instrument of Hooke.

Grew (1628–1711) devoted long and continuous labor to microscopic observation, and, although he was less versatile and brilliant than Hooke, his patient investigations give him just claim to a higher place in the history of natural science. Grew applied the microscope especially to the structure of plants, and his books entitled *Idea of a Philosophical History of Plants* (1673) and *Anatomy of Vegetables* (1682) helped to lay the foundations of vegetable histology. When we come to consider the work of Malpighi, we shall see that he also produced a work upon the microscopic structure of plants which, although not more exact and painstaking than Grew's, showed deeper comprehension. He is the co-founder with Grew of the science of the microscopic anatomy of plants.

It is not necessary to dwell long upon the work of either

Hooke or Grew, since that of Malpighi, Swammerdam, and Leeuwenhoek was more far-reaching in its influence. The publications of these three men were so important, both in reference to microscopic study and to the progress of independent investigation, that it will be necessary to deal with them in more detail. In the work of these men we come upon the first fruits of the application of the methods introduced by Vesalius and Harvey. Of this triumvirate, one—Malpighi—was an Italian, and the other two were Hollanders. Their great service to intellectual progress consisted chiefly in this—that, following upon the foundations of Vesalius and Harvey, "they broke away from the thraldom of mere book-learning, and relying alone upon their own eyes and their own judgment, won for man that which had been quite lost—the blessings of independent and unbiased observation."

It is natural that, working when they did, and independently as they did, their work overlapped in many ways. Malpighi is noteworthy for many discoveries in anatomical science, for his monograph on the anatomy of the silkworm, for observations of the minute structure of plants, and of the development of the chick in the hen's egg. Swammerdam did excellent and accurate work upon the anatomy and metamorphosis of insects, and the internal structure of mollusks, frogs, and other animals. Leeuwenhoek is distinguished for much general microscopic work; he discovered various microscopic animalcula; he established, by direct observation, the fact of a connection between arteries and veins, and examined microscopically minerals, plants, and animals. To him, more than to the others, the general title of "microscopist" might be applied.

Since these men are so important in the growth of biology, let us, by taking them individually, look a little more closely into their lives and labors.

Marcello Malpighi, 1628–1694

Personal Qualities.—There are several portraits of Malpighi extant. These, together with the account of his personal appearance given by Atti, one of his biographers, enable us to tell what manner of man he was. The portrait shown in Fig. 13 is a copy of the one painted by Tabor and presented by Malpighi to the Roval Society of London, in whose rooms it may still be seen. This shows him in the full attractiveness of his early manhood, with the earnest, intellectual look of a man of high ideals and scholarly tastes, sweet-tempered, and endowed with the insight that belongs to a sympathetic nature. Some of his portraits taken later are less attractive, and the lines and wrinkles that show in his face give evidence of imperfect health. According to Atti, he was of medium stature, with a brown skin, a delicate complexion, a serious countenance, and a melancholy look.

Accounts of his life show that he was modest, quiet, and of a pacific disposition, notwithstanding the fact that he lived in an atmosphere of acrimonious criticism, of jealousy and controversy. A familv dispute in reference to the boundary-lines between his father's property and the adjoining land of the Sbaraglia familv gave rise to a feud, in which representatives of the latter family followed him all his life with efforts to injure both his scientific reputation and his good name. Under all this he suffered acutelv, and his removal from Bologna to Messina was partly to escape the harshness of his critics. Some of his best qualities showed under these persecutions; he was dignified under abuse and considerate in his reply. In reference to the attacks upon his scientific standing, there were published after his death replies to his critics that were written while he was smarting under their injustice and severity, but these replies are free from bitterness and are written in a spirit of great moderation. The follow-

FIG. 13.—MALPIGHI, 1628–1694.

ing picture, taken from Ray's correspondence, shows the fine control of his spirit. Under the date of April, 1684, Dr. Tancred Robinson writes: "Just as I left Bononia I had a lamentable spectacle of Malpighi's house all in flames, occasioned by the negligence of his old wife. All his pictures, furniture, books, and manuscripts were burnt. I saw him in the very heat of the calamity, and methought I never beheld so much Christian patience and philosophy in any man before; for he comforted his wife and condoled nothing but the loss of his papers."

Education.—Malpighi was born at Crevalcuore, near Bologna, in 1628. His parents were landed peasants, or farmers, enjoying an independence in financial matters. As their resources permitted it, they designed to give Marcellus, their eldest child, the advantage of masters and schools. He began a life of study; and, before long, he showed a taste for belles-lettres and for philosophy, which he studied under Natali.

Through the death of both parents, in 1649, Malpighi found himself orphaned at the age of twenty-one, and as he was the eldest of eight children, the management of domestic affairs devolved upon him. He had as yet made no choice of a profession; but, through the advice of Natali, he resolved, in 1651, to study medicine. This advice followed, in 1653, at the age of twenty-five, he received from the University of Bologna the degree of Doctor of Medicine.

University Positions.—In the course of a few years he married the sister of Massari, one of his teachers in anatomy, and became a candidate for a chair in the University of Bologna. This he did not immediately receive, but, about 1656, he was appointed to a post in the university, and began his career as a teacher and investigator. He must have shown aptitude for this work, for he was soon called to the University of Pisa, where, fortunately for his development, he became associated with Borelli, who, as an older man, assisted him in many ways. They united in some work, and together they discovered the spiral character of the heart muscles. But the climate of Pisa did not agree with him, and after three years he returned, in 1659, to teach in the University of Bologna, and applied himself assiduously to anatomy.

Here his fame was in the ascendant, notwithstanding the machinations of his enemies and detractors, led by Sbaraglia. He was soon (1662) called to Messina to follow the famous Castelli. After a residence there of four years he again returned to Bologna, and as he was now thirty-eight years of age, he thought it time to retire to his villa near the city in order to devote himself more fully to anatomical studies, but he continued his lectures in the university, and also his practice of medicine.

Honors at Home and Abroad.—Malpighi's talents were appreciated even at home. The University of Bologna honored him in 1686 with a Latin *eulogium;* the city erected a monument to his memory; and after his death, in the city of Rome, his body was brought to Bologna and interred with great pomp and ceremony. At the three hundredth anniversary of his death, in 1894, a festival was held in Bologna, his monument was unveiled, and a book of addresses by eminent anatomists was published in his honor.

During his lifetime he received recognition also from abroad, but that is less remarkable. In 1668 he was elected an honorary member of the Royal Society of London. He was very sensible of this honor; he kept in communication with the society; he presented them with his portrait, and deposited in their archives the original drawings illustrating the anatomy of the silkworm and the development of the chick.

In 1691 he was taken to Rome by the newly elected pope, Innocent XII, as his personal physician, but under these new conditions he was not destined to live many years. He died there, in 1694, of apoplexy. His wife, of whom it appears that he was very fond, had died a short time previously. Among his posthumous works is a sort of personal psychology written down to the year 1691, in which he shows the growth of his mind, and the way in which he came to take up the different subjects of investigation.

In reference to his discoveries and the position he occupies in the history of natural science, it should be observed that he was an "original as well as a very profound observer." While the ideas of anatomy were still vague, "he applied himself with ardor and sagacity to the study of the fine structure of the different parts of the body," and he extended his investigations to the structure of plants and of different animals, and also to their development. Entering, as he did, a new and unexplored territory, naturally he made many discoveries, but no man of mean talents could have done his work.

Activity in Research.—During forty years of his life he was always busy with research. Many of his discoveries had practical bearing on the advance of anatomy and physiology as related to medicine. In 1661 he demonstrated the structure of the lungs. Previously these organs had been regarded as a sort of homogeneous parenchyma. He showed the presence of air-cells, and had a tolerably correct idea of how the air and the blood are brought together in the lungs, the two never actually in contact, but always separated by a membrane. These discoveries were first made on the frog, and applied by analogy to the interpretation of the lungs of the human body. He was a comparative anatomist, and the first to insist on analogies of structure between organs throughout the animal kingdom, and to make extensive practical use of the idea that discoveries on simpler animals can be utilized in interpreting the similar structures in the higher ones.

It is very interesting to note that in connection with this work he actually observed the passage of blood through the capillaries of the transparent lungs of the frog, and also in the mesentery. Although this antedates the similar observations of Leeuwenhoek (1669), nevertheless the work of Leeuwenhoek was much more complete, and he is usually recognized in physiology as the discoverer of the capillary connection between arteries and veins. At this same period Malpighi also observed the blood corpuscles.

Soon after he demonstrated the mucous layer, or pigmentary layer of the skin, intermediate between the true and the scarf skin. He had separated this layer by boiling and maceration, and described it as a reticulated membrane. Even its existence was for a long time controverted, but it remains in modern anatomy under the title of the Malpighian layer.

His observation of glands was extensive, and while it must be confessed that many of his conclusions in reference to glandular structure were erroneous, he left his name connected with the Malpighian corpuscles of the kidney and of the spleen. He was also the first to indicate the nature of the papillæ on the tongue. The foregoing is a respectable list of discoveries, but much more stands to his credit. Those which follow have a bearing on comparative anatomy, zoölogy, and botany.

Monograph on the Structure and Metamorphosis of the Silkworm.—Malpighi's work on the structure of the silkworm takes rank among the most famous monographs on the anatomy of a single animal. Much skill was required to give to the world this picture of minute structure. The marvels of organic architecture were being made known in the human body and the higher animals, but "no insect—hardly, indeed, any animal—had then been carefully described, and all the methods of the work had to be discovered." He labored with such enthusiasm in this new territory as to throw himself into a fever and to set up an inflammation in the eyes. "Nevertheless," says Malpighi, "in performing these researches so many marvels of nature were spread before my eyes that I experienced an internal pleasure that my pen could not describe."

He showed that the method of breathing was neither by lungs nor by gills, but through a system of air-tubes, communicating with the exterior through buttonhole-shaped openings, and, internally, by an infinitude of branches reaching to the minutest parts of the body. Malpighi showed an instinct for comparison; instead of confining his researches to the species in hand, he extended his observations to other insects, and has given sketches of the breathing-tubes, held open by their spiral thread, taken from several species.

The nervous system he found to be a central white cord with swellings in each ring of the body, from which nerves are given off to all organs and tissues. The cord, which is, of course, the central nervous system, he found located mainly on the ventral surface of the body, but extending by a sort of collar of nervous matter around the œsophagus, and on the dorsal surface appearing as a more complex mass, or brain, from which nerves are given off to the eyes and other sense organs of the head. As illustrations from this monograph we have, in Fig. 14, reduced sketches of the drawings of the nervous system and the food canal in the adult silkworm. The sketch at the right hand illustrates the central nerve cord with its ganglionic enlargement in each segment, the segments being indicated by the rows of spiracles at the sides. The original drawing is on a much larger scale, and reducing it takes away some of its coarseness. All of his drawings lack the finish and detail of Swammerdam's work.

He showed also the food canal and the tubules connected

with the intestine, which retain his name in the insect anatomy of to-day, under the designation of Malpighian tubes. The silk-forming apparatus was also figured and described. These

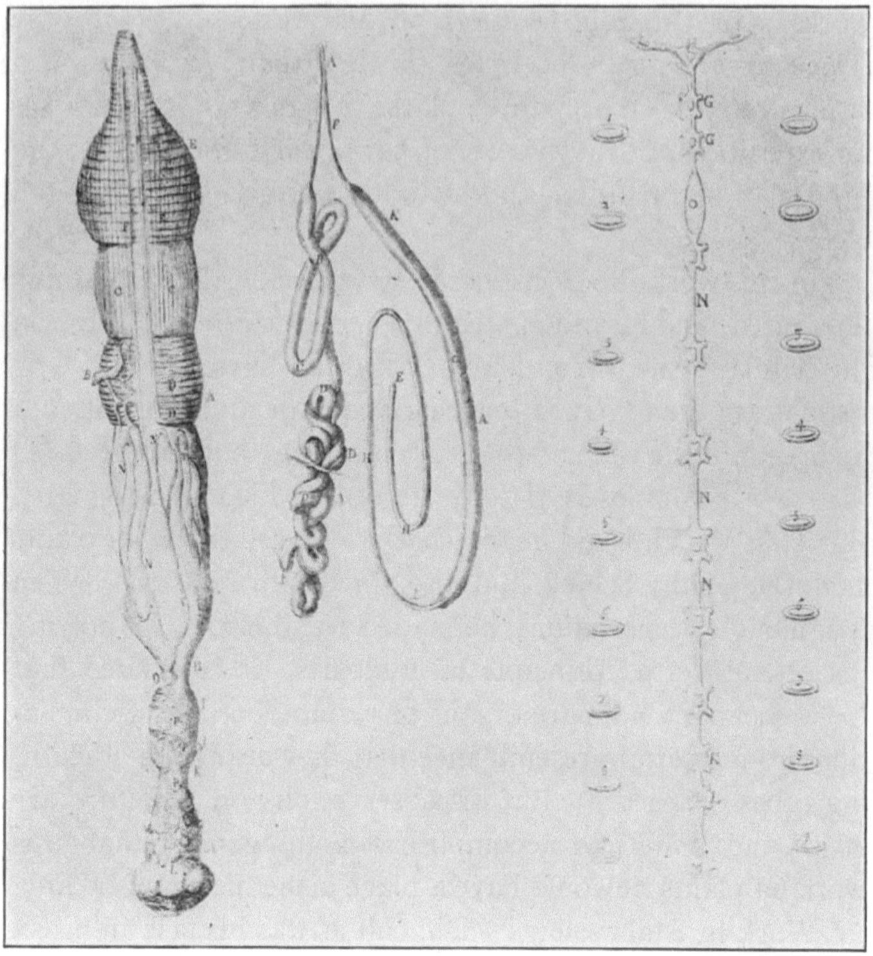

FIG. 14.—From Malpighi's *Anatomy of the Silkworm*, 1669.

structures are represented, as Malpighi drew them, on the left of Fig. 14.

This monograph, which was originally published in 1669 by the Royal Society of London, bears the Latin title, *Dissertatio Epistolica de Bombyce.* It has been several times republished, the best edition being that in French, which dates from Montpellier, in 1878, and which is prefaced by an account of the life and labors of Malpighi.

Anatomy of Plants.—Malpighi's anatomy of plants constitutes one of his best, as well as one of his most extensive works. In the folio edition of his works, 1675–79, the *Anatome Plantarum* occupies not less than 152 pages and is illustrated by ninety-three plates of figures. It comprises an exposition of the structure of bark, stem, roots, seeds, the process of germination, and includes a treatise on galls, etc., etc.

In this work the microscopic structure of plants is amply illustrated, and he anticipated to a certain degree the ideas on the cellular structure of plants. Burnett says: "His observations appear to have been very accurate, and not only did he maintain the cellular structure of plants, but also declared that it was composed of separate cells, which he designated 'utricles.'" Thus did he foreshadow the cell theory of plants as developed by Schleiden in the nineteenth century. When it came to interpretations, he made several errors. Applying his often-asserted principle of analogies, he concluded that the vessels of plants are organs of respiration and of circulation, from a certain resemblance that they bear to the breathing-tubes of insects. But his observations on structure are good, and if he had accomplished nothing more than this work on plants he would have a place in the history of botany.

Work in Embryology.—Difficult as was his task in insect anatomy and plant histology, a more difficult one remains to be mentioned, *viz.*, his observations of the development of animals. He had pushed his researches into the finer structure of organisms, and now he attempted to answer this question: How does one of these organisms begin its life, and by what series of steps is its body built up? He turned to the chick, as the most available form in which to get an insight into this process, but he could not extend his observations successfully into periods earlier than about the twenty-four-hour stage of development. Two memoirs were written on this subject, both in 1672, which were published by the Royal Society of England under the titles *De Formatione Pulli in Ovo* and *De Ovo Incubato.* Of all Malpighi's work, this has received the least attention from reviewers, but it is, for his time, a very remarkable achievement. No one can look over the ten folio plates without being impressed with the extent and accuracy of his observations. His sketches are of interest, not only to students of embryology, but also to educated people, to see how far observations regarding the development of animals had progressed in 1672. Further consideration of his position in embryology will be found in the chapter on the rise of that subject.

Little is known regarding the form of microscope employed by Malpighi. Doubtless, much of his work was done with a simple lens, since he speaks of examining the dried lungs with a microscope of a single lens against the horizontal sun; but he is also known to have observed with an instrument consisting of two lenses.

Malpighi was a naturalist, but of a new type; he began to look below the surface, and essayed a deeper level of analysis in observing and describing the internal and minute structure of animals and plants, and when he took the further step of investigating their development he was anticipating the work of the nineteenth century.

JAN SWAMMERDAM (1637–1680)

Swammerdam was a different type of man—nervous, incisive, very intense, stubborn, and self-willed. Much of his character shows in the portrait by Rembrandt represented in Fig. 15. Although its authenticity has been questioned, it is the only known portrait of Swammerdam.

Early Interest in Natural History.—He was born in 1637, nine years after Malpighi. His father, an apothecary of Amsterdam, had a taste for collecting, which was shared by many of his fellow-townsmen. The Dutch people of this time sent their ships into all parts of the world, and this vast commerce, together with their extensive colonial possessions, fostered the formation of private museums. The elder Swammerdam had the finest and most celebrated collection in all Amsterdam. This was stored, not only with treasures, showing the civilization of remote countries, but also with specimens of natural history, for which he had a decided liking. Thus "from the earliest dawn of his understanding the young Swammerdam was surrounded by zoölogical specimens, and from the joint influence, doubtless, of hereditary taste and early association, he became passionately devoted to the study of natural history."

Studies Medicine.—His father intended him for the church, but he had no taste for theology, though he became a fanatic in religious matters toward the close of his life; at this period, however, he could brook no restraint in word or action. He consented to study medicine, but for some reason he was twenty-six years old before entering the University of Leyden. This delay was very likely owing to his precarious health, but, in the mean time, he had not been idle; he had devoted himself to observation and study with great ardor, and had already become an expert in minute dissection. When he went to the University of Leyden, therefore, he at once took high rank in anatomy. Anything demanding fine manipulation and dexterity was directly in his line. He continued his studies in Paris, and about 1667 took his degree of Doctor of Medicine.

During this period of medical study he made some rather important observations in human anatomy, and introduced the method of injection that was afterward claimed by

FIG. 15.—SWAMMERDAM, 1637–1680.

Ruysch. In 1664 he discovered the valves of lymphatic vessels by the use of slender glass tubes, and, three years later, first used a waxy material for injecting blood-vessels.

It should be noted, in passing, that Swammerdam was the first to observe and describe the blood corpuscles. As early as 1658 he described them in the blood of the frog, but not till fifty-seven years after his death were his observations published by Boerhaave, and, therefore, he does not get the credit of this discovery. Publication alone, not first observation, establishes priority, but there is conclusive evidence that he observed the blood corpuscles before either Malpighi or Leeuwenhoek had published his findings.

Love of Minute Anatomy.—After graduating in medicine he did not practice, but followed his strong inclination to devote himself to minute anatomy. This led to differences with his father, who insisted on his going into practice, but the self-willed stubbornness and firmness of the son now showed themselves. It was to gratify no love of ease that Swammerdam thus held out against his father, but to be able to follow an irresistible leading toward minute anatomy. At last his father planned to stop supplies, in order to force him into the desired channel, but Swammerdam made efforts, without success, to sell his own personal collection and preserve his independence. His father died, leaving him sufficient property to live on, and brought the controversy to a close soon after the son had consented to yield to his wishes.

Boerhaave, his fellow-countryman, gathered Swammerdam's complete writings after his death and published them in 1737 under the title *Biblia Naturæ*. With them is included a life of Swammerdam, in which a graphic account is given of his phenomenal industry, his intense application, his methods and instruments. Most of the following passages are selected from that work.

Intensity as a Worker.—He was a very intemperate

worker, and in finishing his treatise on bees (1673) he broke himself down.

"It was an undertaking too great for the strongest constitution to be continually employed by day in making observations and almost as constantly engaged by night in recording them by drawings and suitable explanations. This being summer work, his daily labors began at six in the morning, when the sun afforded him light enough to enable him to survey such minute objects; and from that time till twelve he continued without interruption, all the while exposed in the open air to the scorching heat of the sun, bareheaded, for fear of interrupting the light, and his head in a manner dissolving into sweat under the irresistible ardors of that powerful luminary. And if he desisted at noon, it was only because the strength of his eyes was too much weakened by the extraordinary efflux of light and the use of microscopes to continue any longer upon such small objects.

"This fatigue our author submitted to for a whole month together, without any interruption, merely to examine, describe, and represent the intestines of bees, besides many months more bestowed upon the other parts; during which time he spent whole days in making observations, as long as there was sufficient light to make any, and whole nights in registering his observations, till at last he brought his treatise on bees to the wished-for perfection."

Method of Work.—"For dissecting very minute objects, he had a brass table made on purpose by that ingenious artist, Samuel Musschenbroek. To this table were fastened two brass arms, movable at pleasure to any part of it, and the upper portion of these arms was likewise so contrived as to be susceptible of a very slow vertical motion, by which means the operator could readily alter their height as he saw most convenient to his purpose. The office of one of these arms was to hold the little corpuscles, and that of the other to apply the microscope. His microscopes were of various sizes and curvatures, his microscopical glasses being of various diameters and focuses, and, from the least to the greatest, the best that could be procured, in regard to the exactness of the workmanship and the transparency of the substance.

"But the constructing of very fine scissors, and giving them an extreme sharpness, seems to have been his chief secret. These he made use of to cut very minute objects, because they dissected them equably, whereas knives and lancets, let them be ever so fine and sharp, are apt to disorder delicate substances. His knives, lancets, and styles were so fine that he could not see to sharpen them without the assistance of the microscope; but with them he could dissect the intestines of bees with the same accuracy and distinctness that others do those of large animals.

"He was particularly dexterous in the management of small tubes of glass no thicker than a bristle, drawn to a very fine point at one end, but thicker at the other."

These were used for inflating hollow structures, and also for making fine injections. He dissolved the fat of insects in turpentine and carried on dissections under water.

An unbiased examination of his work will show that it is of a higher quality than Malpighi's in regard to critical observation and richness of detail. He also worked with minuter objects and displayed a greater skill.

The Religious Devotee.—The last part of his life was dimmed by fanaticism. He read the works of Antoinette Bourignon and fell under her influence; he began to subdue his warm and stubborn temper, and to give himself up to religious contemplation. She taught him to regard scientific research as worldly, and, following her advice, he gave up his passionate fondness for studying the works of the Creator, to devote himself to the love and adoration of that same Being. Always extreme and intense in everything he undertook, he likewise overdid this, and yielded himself to a sort of fanatical worship until the end of his life, in 1680. Had he possessed a more vigorous constitution he would have been greater as a man. He lived, in all, but fortv-three years; the last six or seven years were unproductive because of his mental distractions, and before that, much of his time had been lost through sickness.

The Biblia Naturæ.—It is time to ask, What, with all his talents and prodigious application, did he leave to science? This is best answered by an examination of the *Biblia Naturæ*, under which title all his work was collected. His treatise on Bees and Mayflies and a few other articles were published during his lifetime, but a large part of his observations remained entirelv unknown until they were published in this book fifty-seven years after his death. In the folio edition it embraces 410 pages of text and fifty-three plates, replete with figures of original observations. It "contains about a dozen life-histories of insects worked out in more or less detail. Of these, the mayfly is the most famous; that on the honey-bee the most elaborate." The greater amount of his work was in structural entomology. It is known that he had a collection of about three thousand different species of insects, which for that period was a very large one. There is, however, a considerable amount of work on other animals; the fine anatomy of the snail, the structure of the clam, the squid; observations on the structure and development of the frog; observations on the contraction of the muscles, etc., etc.

It is to be remembered that Swammerdam was extremely exact in all that he did. His descriptions are models of accuracy and completeness.

Fig. 16 shows reduced sketches of his illustrations of the structure of the snail. The upper sketch shows the central nervous system and the nerve trunks connected therewith, and the lower figure shows the shell and the principal muscles.

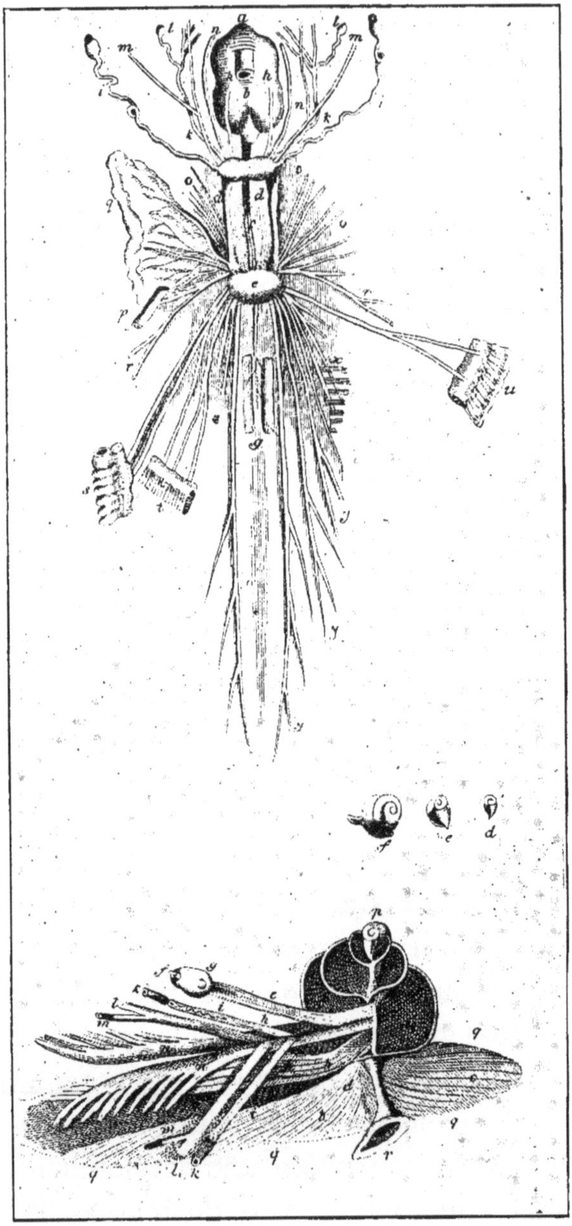

FIG. 16.—From Swammerdam's *Biblia Naturæ*.

This is an exceptionally good piece of anatomization for that time, and is a fair sample of the fidelity with which he worked out details in the structure of small animals. Besides showing this, these figures also serve the purpose of pointing out that Swammerdam's fine anatomical work was by no means confined to insects. His determinations on the structure of the young frog were equally noteworthy.

But we should have at least one illustration of his handling of insect anatomy to compare more directly with that of Malpighi, already given. Fig. 17 is a reduced sketch of the anatomy of the larva of an ephemerus, showing, besides other structures, the central nervous system in its natural position. When compared with the drawings of Malpighi, we see there is a more masterly hand at the task, and a more critical spirit back of the hand. The nervous system is very well done, and the greater detail in other features shows a disposition to go into the subject more deeply than Malpighi.

Besides working on the structure and life-histories of animals, Swammerdam showed, experimentally, the irritability of nerves and the response of muscles after their removal from the body. He not only illustrates this quite fully, but seems to have had a pretty good appreciation of the nature of the problem of the physiologist. He says:

"It is evident from the foregoing observations that a great number of things concur in the contraction of the muscles, and that one should be thoroughly acquainted with that wonderful machine, our body, and the elements with which we are surrounded, to describe exactly one single muscle and explain its action. On this occasion it would be necessary for us to consider the atmosphere, the nature of our food, the blood, the brain, marrow, and nerves, that most subtle matter which instantaneously flows to the fibers, and many other things, before we could expect to attain a sight of the perfect and certain truth."

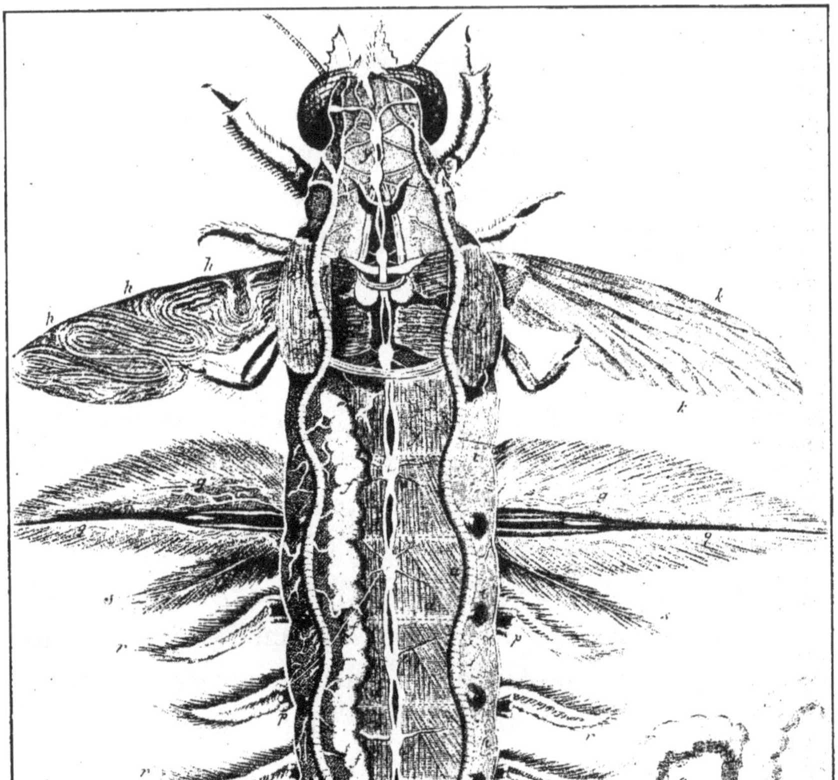

Fig. 17.—Anatomy of an Insect: Dissected and Drawn by Swammerdam.

In reference to the formation of animals within the egg, Swammerdam was, as Malpighi, a believer in the pre-formation theory. The basis for his position on this question will be set forth in the chapter on the Rise of Embryology.

There was another question in his time upon which philosophers and scientific men were divided, which was in reference to the origin of living organisms: Does lifeless matter, sometimes, when submitted to heat and moisture, spring into life? Did the rats of Egypt come, as the ancients believed, from the mud of the Nile, and do frogs and toads have a similar origin? Do insects spring from the dew on plants? etc., etc. The famous Redi performed his noteworthy experiments when Swammerdam was twenty-eight years old, but opinion was divided upon the question as to the possible spontaneous origin of life, especially among the smaller animals. Upon this question Swammerdam took a positive stand; he ranged himself on the side of the more scientific naturalists against the spontaneous formation of life.

Antony van Leeuwenhoek (1632–1723)

In Leeuwenhoek we find a composed and better-balanced man. Blessed with a vigorous constitution, he lived ninety-one years, and worked to the end of his life. He was born in 1632, four years after Malpighi, and five before Swammerdam; they were, then, strictly speaking, contemporaries. He stands in contrast with the other men in being self-taught; he did not have the advantage of a university training, and apparently never had a master in scientific study. This lack of systematic training shows in the desultory character of his extensive observations. Impelled by the same gift of genius that drove his confrères to study nature with such unexampled activity, he too followed the path of an independent and enthusiastic investigator.

The portrait (Fig. 18) which forms a frontispiece to his *Arcana Naturæ* represents him at the age of sixty-three, and shows the pleasing countenance of a firm man in vigorous health. Richardson says: "In the face peering through the big wig there is the quiet force of Cromwell and the delicate disdain of Spinoza." "It is a mixed racial type, Semitic and Teutonic, a Jewish-Saxon; obstinate and yet imaginative; its very obstinacy a virtue, saving it from flying too far wild by its imagination."

Recent Additions to His Biography.—There was a singular scarcity of facts in reference to Leeuwenhoek's life until 1885, when Dr. Richardson published in *The Asclepiad* * the results of researches made by Mr. A. Wynter Blyth in Leeuwenhoek's native town of Delft. I am indebted to that article for much that follows.

His *Arcana Naturæ* and other scientific letters contained a complete record of his scientific activity, but "about his parentage, his education, and his manner of making a living there was nothing but conjecture to go upon." The few scraps of personal history were contained in the Encyclopædia articles by Carpenter and others, and these were wrong in sustaining the hypothesis that Leeuwenhoek was an optician or manufacturer of lenses for the market. Although he ground lenses for his own use, there was no need on his part of increasing his financial resources by their sale. He held under the court a minor office designated 'Chamberlain of the Sheriff.' The duties of the office were those of a beadle, and were set forth in his commission, a document still extant. The requirements were light, as was also the salary, which amounted to about £26 a year. He held this post for thirty-nine years, and the stipend was thereafter continued to him to the end of his life.

Van Leeuwenhoek was derived from a good Delft family.

* *Leeuwenhoek and the Rise of Histology.* The Asclepiad, Vol. II, 1885.

FIG. 18.—LEEUWENHOEK, 1632–1723.

His grandfather and his great-grandfather were Delft brewers, and his grandmother a brewer's daughter. The family were doubtless wealthy. His schooling seems to have been brought to a close at the age of sixteen, when he was "removed to a clothing business in Amsterdam, where he filled the office of bookkeeper and cashier." After a few years he returned to Delft, and at the age of twenty-two he married, and gave himself up largely to studies in natural history. Six years after his marriage he obtained the appointment mentioned above. He was twice married, but left only one child, a daughter by his first wife. In the old church at Delft is a monument erected by this daughter to the memory of her father.

He led an easy, prosperous, but withal a busy life. The microscope had recently been invented, and for observation with that new instrument Leeuwenhoek showed an avidity amounting to a passion.

"That he was in comfortable, if not affluent, circumstances is clear from the character of his writings; that he was not troubled by any very anxious and responsible duties is certain from the continuity of his scientific work; that he could secure the services of persons of influence is discernible from the circumstances that, in 1673, De Graaf sent his first paper to the Royal Society of London; that in 1680 the same society admitted him as fellow; that the directors of the East India Company sent him specimens of natural history, and that, in 1698, Peter the Great paid him a call to inspect his microscopes and their revelations."

Leeuwenhoek seems to have been fascinated by the marvels of the microscopic world, but the extent and quality of his work lifted him above the level of the dilettante. He was not, like Malpighi and Swammerdam, a skilled dissector, but turned his microscope in all directions; to the mineral as well as to the vegetable and animal kingdoms. Just when he began to use the microscope is not known; his first publication in reference to microscopic objects did not appear till 1673, when he was forty-one years old.

His Microscopes.—He gave good descriptions and drawings of his instruments, and those still in existence have been described by Carpenter and others, and in consequence we have a very good idea of his working equipment. During his lifetime he sent as a present to the Royal Society of London twenty-six microscopes, each provided with an object to examine. Unfortunately, these were removed from the rooms of the society and lost during the eighteenth century. His lenses were of fine quality and were ground by himself. They were nearly all simple lenses, of small size but considerable curvature, and needed to be brought close to the object examined. He had different microscopes for different purposes, giving a range of magnifying powers from 40 to 270 diameters and possibly higher. The number of his lenses is surprising; he possessed not less than 247 complete microscopes, two of which were provided with double lenses, and one with a triplet. In addition to the above, he had 172 lenses set between plates of metal, which give a total of 419 lenses used by him in his observations. Three were of quartz, or rock crystal; the rest were of glass. More than one-half the lenses were mounted in silver; three were in gold.

It is to be understood that all his microscopes were of simple construction; no tubes, no mirror; simple pieces of metal to hold the magnifying-glass and the objects to be examined, with screws to adjust the position and the focus.

The three aspects of one of Leeuwenhoek's microscopes shown in Fig. 19 will give a very good idea of how they were constructed. These pictures represent the actual size of the instrument. The photographs were made by Professor

Nierstrasz from the specimen in possession of the University of Utrecht. The instrument consists of a double copper plate in which the circular lens is inserted, and an object-holder—represented in the right-hand lower figure as thrown to one

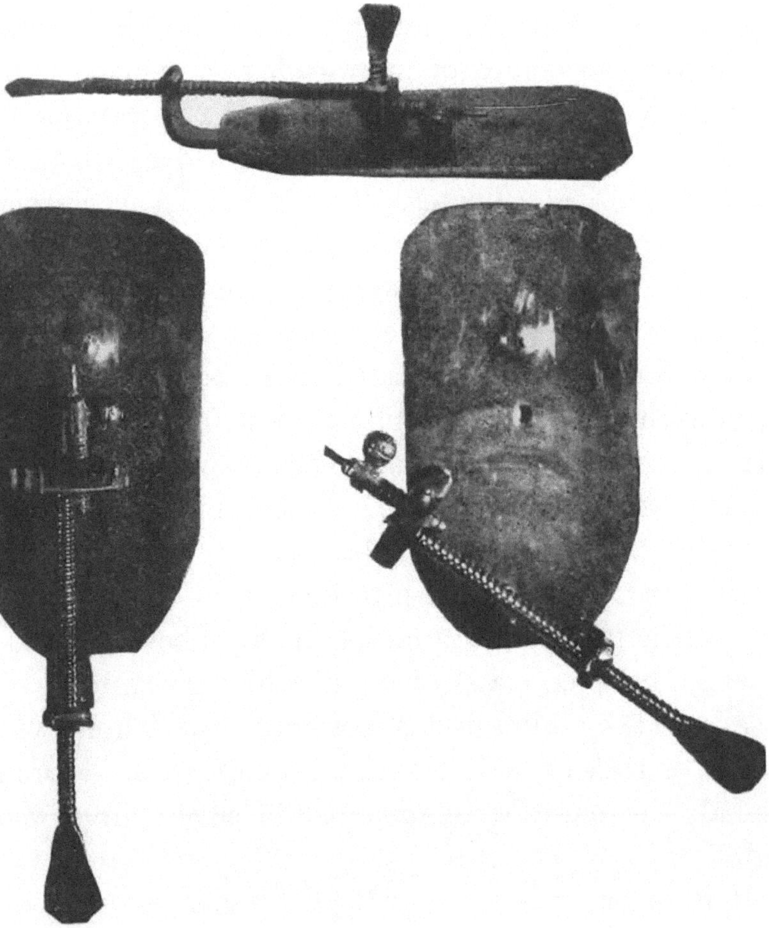

FIG. 19.—LEEUWENHOEK'S MICROSCOPE.
Natural size. From Photographs by Professor Nierstrasz, of Utrecht.

side. By a vertical screw the object could be elevated or depressed, and by a transverse screw it could be brought nearer or removed farther from the lens, and thus be brought into focus.

Fig. 20*a* shows the way in which the microscope was

arranged to examine the circulation of blood in the transparent tail of a small fish. The fish was placed in water in a slender glass tube, and the latter was held in a metallic frame, to which a plate (marked *D*) was joined, carrying the magnifying glass. The latter is indicated in the circle above the letter *D*, near the tail-fin of the fish. The eye was applied close to this circular magnifying-glass, which was brought into position and adjusted by means of screws. In some instances, he had a concave reflector with a hole in the center, in which his magnifying-glass was inserted; in this form of instrument the objects were illumined by reflected, and not by transmitted light.

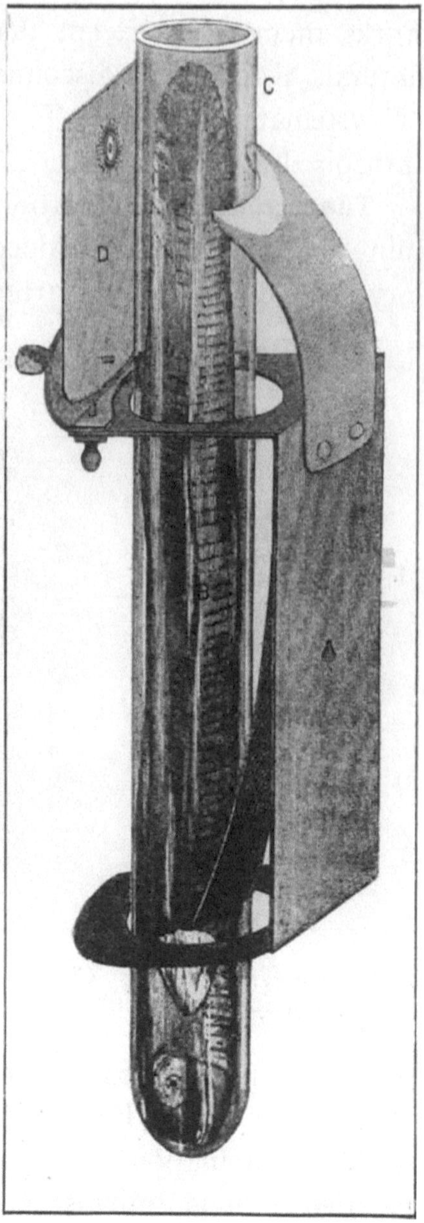

FIG. 20*a*. — Leeuwenhoek's Mechanism for Examining the Circulation of the Blood.

His Scientific Letters.— His microscopic observations were described and sent to learned societies in the form of letters. "All or nearly all that he did in a literary way was after the manner of an epistle," and his written communications were so numerous as to justify the cognomen, "The man of many letters." "The French Academy of Sciences, of which he was elected a corresponding member in 1697, got twenty-seven; but the lion's share fell to the young Royal Society of London, which in fifty years—1673–1723—received 375 letters and papers." "The works themselves, except that they lie in the domain of natural history, are disconnected and appear in no order of systematized study. The philosopher was led by what transpired at any moment to lead him."

The Capillary Circulation.—In 1686 he observed the minute circulation of the blood, and demonstrated the capillary connection between arteries and veins, thus forging the final link in the chain of observation showing the relation between these blood-vessels. This was perhaps his most important observation for its bearing on physiology. It must be remembered that Harvey had not actually seen the circulation of the blood, which he announced in 1628. He assumed on entirely sufficient grounds the existence of a complete cir culation, but there was wanting in his scheme the direct ocular proof of the passage of blood from arteries to veins. This was supplied by Leeuwenhoek. Fig. 20*b* shows one of his sketches of the capillary circulation. In his efforts to see the circulation he tried various animals; the comb of the young cock, the ears of white rabbits, the membraneous wing of the bat were progressively examined. The next advance came when he

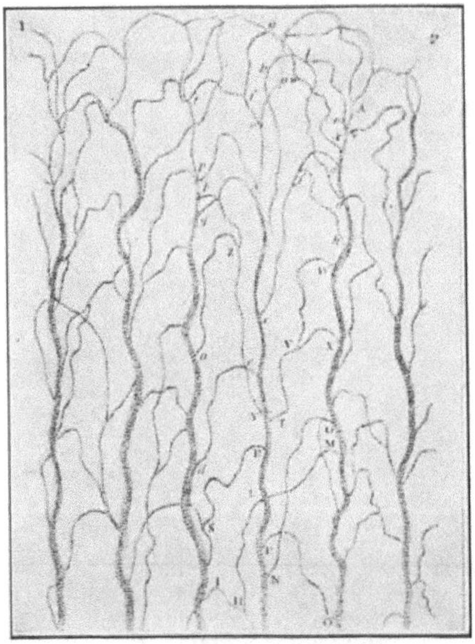

FIG. 20*b*.—The Capillary Circulation. (After Leeuwenhoek.)

directed his microscope to the tail of the tadpole. Upon examining this he exclaims:

"A sight presented itself more delightful than any mine eyes had ever beheld; for here I discovered more than fifty circulations of the blood in different places, while the animal lay quiet in the water, and I could bring it before my microscope to my wish. For I saw not only that in many places the blood was conveyed through exceedingly minute vessels, from the middle of the tail toward the edges, but that each of the vessels had a curve or turning, and carried the blood back toward the middle of the tail, in order to be again conveyed to the heart. Hereby it plainly appeared to me that the blood-vessels which I now saw in the animal, and which bear the names of arteries and veins are, in fact, one and the same; that is to say, that they are properly termed arteries so long as they convey the blood to the furtherest extremities of its vessels, and veins when they bring it back to the heart. And thus it appears that an artery and a vein are one and the same vessel prolonged or extended."

This description shows that he fully appreciated the course of the minute vascular circulation and the nature of the communication between arteries and veins. He afterward extended his observations to the web of the frog's foot, the tail of young fishes and eels.

In connection with this it should be remembered that Malpighi, in 1661, observed the flow of blood in the lungs and in the mesentery of the frog, but he made little of the discovery. Leeuwenhoek did more with his, and gave the first clear idea of the capillary circulation. Leeuwenhoek was anticipated also by Malpighi in reference to the microscopic structure of the blood. (See also under Swammerdam.) To Malpighi the corpuscles appeared to be globules of fat, while Leeuwenhoek noted that the blood disks of birds, frogs, and fishes were oval in outline, and those of mammals circular. He reserved the term 'globule' for those of the human body, erroneously believing them to be spheroidal.

Other Discoveries.—Among his other discoveries bearing on physiology and medicine may be mentioned: the branched character of heart muscles, the stripe in voluntary muscles, the structure of the crystalline lens, the description of spermatozoa after they had been pointed out to him in 1674 by Hamen, a medical student in Leyden, etc. Richardson dignified him with the title 'the founder of histology,' but this, in view of the work of his great contemporary, Malpighi, seems to me an overestimate.

Turning his microscope in all directions, he examined water and found it peopled with minute animalcules, those simple forms of animal life propelled through the water by innumerable hair-like cilia extending from the body like banks of oars from a galley, except that in many cases they extend from all surfaces. He saw not only the animalcules, but also the cilia that move their bodies.

He also discovered the Rotifers, those favorites of the amateur microscopists, made so familiar to the general public in works like Gosse's *Evenings at the Microscope.* He observed that when water containing these animalcules evaporated they were reduced to fine dust, but became alive again, after great lapses of time, by the introduction of water.

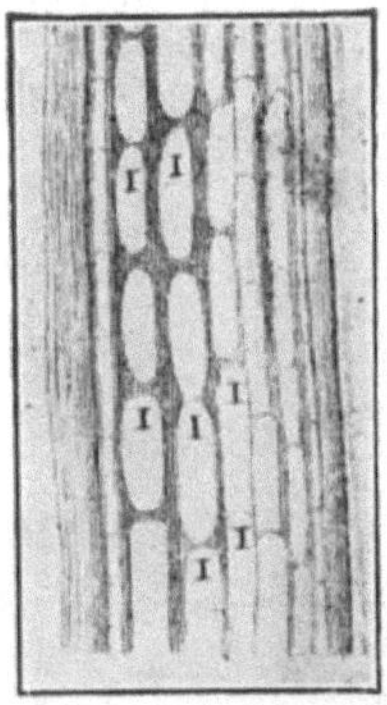

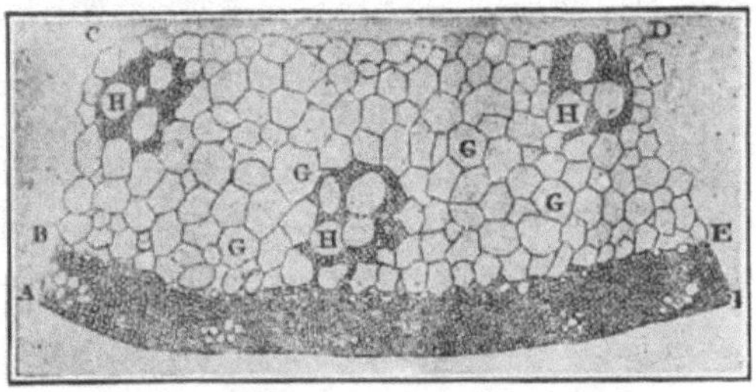

FIG. 21.—Plant Cells. (From Leeuwenhoek's *Arcana Naturæ.*)

He made many observations on the

microscopic structure of plants. Fig. 21 gives a fair sample of the extent to which he observed the cellular construction of vegetables and anticipated the cell theory. While Malpighi's research in that field was more extensive, these sketches from Leeuwenhoek represent very well the character of the work of the period on the minute structures of plants.

His Theoretical Views.—It remains to say that on the two biological questions of the day he took a decisive stand. He was a believer in pre-formation or pre-delineation of the embryo in an extreme degree, seeing in fancy the complete outline of both maternal and paternal individuals in the spermatozoa, and going so far as to make sketches of the same. But on the question of the spontaneous origin of life he took the side that has been supported with such triumphant demonstration in this century; namely, the side opposing the theory of the occurrence of spontaneous generation under present conditions of life.

Comparison of the Three Men.—We see in these three gifted contemporaries different personal characteristics. Leeuwenhoek, the composed and strong, attaining an age of ninety-one; Malpighi, always in feeble health, but directing his energies with rare capacity, reaching the age of sixty-seven; while the great intensity of Swammerdam stopped his scientific career at thirty-six and burned out his life at the age of forty-three.

They were all original and accurate observers, but there is variation in the kind and quality of their intellectual product. The two university-trained men showed capacity for coherent observation; they were both better able to direct their efforts toward some definite end; Leeuwenhoek, with the advantages of vigorous health and long working period, lacked the systematic training of the schools, and all his life wrought in discursive fashion; he left no coherent piece of work of any extent like Malpighi's *Anatome Plantarum* or Swammerdam's *Anatomy and Metamorphosis of Insects.*

Swammerdam was the most critical observer of the three, if we may judge by his labors in the same field as Malpighi's on the silkworm. His descriptions are models of accuracy and completeness, and his anatomical work shows a higher grade of finish and completeness than Malpighi's. Malpighi, it seems to me, did more in the sum total than either of the others to advance the sciences of anatomy and physiology, and through them the interests of mankind. Leeuwenhoek had larger opportunity; he devoted himself to microscopic observations, but he wandered over the whole field. While his observations lose all monographic character, nevertheless they were important in opening new fields and advancing the sciences of anatomy, physiology, botany, and zoölogy.

The combined force of their labors marks an epoch characterized by the acceptance of the scientific method and the establishment of a new grade of intellectual life. Through their efforts and that of their contemporaries of lesser note the new intellectual movement was now well under way.

CHAPTER V

THE PROGRESS OF MINUTE ANATOMY.

THE work of Malpighi, Swammerdam, and Leeuwenhoek stimulated investigations into the structure of minute animals, and researches in that field became a feature of the advance in the next century. Considerable progress was made in the province of minute anatomy before comparative anatomy grew into an independent subject.

The attractiveness of observations upon the life-histories and the structure of insects, as shown particularly in the publications of Malpighi and Swammerdam, made those animals the favorite objects of study. The observers were not long in recognizing that some of the greatest beauties of organic architecture are displayed in the internal structure of insects. The delicate tracery of the organs, their minuteness and perfection are well calculated to awaken surprise. Well might those early anatomists be moved to enthusiasm over their researches. Every excursion into this domain gave only beautiful pictures of a mechanism of exquisite delicacy, and their wonder grew into amazement. Here began a new train of ideas, in the unexpected revelation that within the small compass of the bodv of an insect was embraced such a complex set of organs; a complete nervous system, fine breathing-tubes, organs of circulation, of digestion, etc., etc.

Lyonet.—The first piece of structural work after Swammerdam's to which we must give attention is that of Lyonet, who produced in the middle of the eighteenth century one of the most noteworthy monographs in the field of minute anatomy. This was a work like that of Malpighi, upon the anatomy of a single form, but it was carried out in much

FIG. 22.—LYONET, 1707–1789.

greater detail. The 137 figures on the 18 plates are models of close observation and fine execution of drawings.

Lyonet (also written Lyonnet) was a Hollander, born in The Hague in 1707. He was a man of varied talents, a painter, a sculptor, an engraver, and a very gifted linguist.

It is said that he was skilled in at least eight languages; and at one time he was the cipher secretary and confidential translator for the United Provinces of Holland. He was educated as a lawyer, but, from interest in the subject, devoted most of his time to engraving objects of natural history. Among his earliest published drawings were the figures for Lesser's *Theology of Insects* (1742) and for Trembley's famous treatise on *Hydra* (1744).

His Great Monograph.—Finally Lyonet decided to branch out for himself, and produce a monograph on insect anatomy. After some preliminary work on the sheep-tick, he settled upon the caterpillar of the goat moth, which lives upon the willow-tree. His work, first published in 1750, bore the title *Traité Anatomique de la Chenille qui ronge le bois de Saule.* In exploring the anatomy of the form chosen, he displayed not only patience, but great skill as a dissector, while his superiority as a draughtsman was continually shown in his sketches. He engraved his own figures on copper. The drawings are very remarkable for the amount of detail that they show. He dissected this form with the same thoroughness with which medical men have dissected the human body. The superficial muscles were carefully drawn and were then cut away in order to expose the next underlying layer which, in turn, was sketched and then removed. The amount of detail involved in this work may be in part realized from the circumstance that he distinguished 4,041 separate muscles. His sketches show these muscles accurately drawn, and the principal ones are lettered. When he came to expose the nerves, he followed the minute branches to individual small muscles and sketched them, not in a diagrammatic way, but as accurate drawings from the natural object. The breathing-tubes were followed in the same manner, and the other organs of the body were all dissected and drawn with remarkable thoroughness. Lyonet was not trained in anatomy like Malpighi and Swammerdam, but being a man of unusual patience and manual dexterity, he accomplished notable results. His great quarto volume is, however, merely a description of the figures, and lacks the insight of a trained anatomist. His skill as a dissector is far ahead of his knowledge of anatomy, and he becomes lost in the details of his subject.

FIG. 23.—Larva of the Willow Moth. (From Lyonet's Monograph, 1750.)

Extraordinary Quality of the Drawings.—A few figures will serve to illustrate the character of his work, but the reduced reproductions which follow can not do justice to the copper plates of the original. Fig. 23 gives a view of the external appearance of the caterpillar which was dissected. When the skin was removed from the outside the muscles came into view, as shown in Fig. 24. This is a view from the ventral side of the animal. On the left side the more superficial muscles show, and on the right the next deeper layer.

Fig. 25 shows his dissection of the nerves. In this figure the muscles are indicated in outline, and the distribution of nerves to partieular muscles is shown.

Lyonet's dissection of the head is an extraordinary feat. The entire head is not more than a quarter of an inch in diameter, but in a series of seven dissections he shows all of the internal organs in the head. Fig. 26 shows two sketches

FIG. 24.

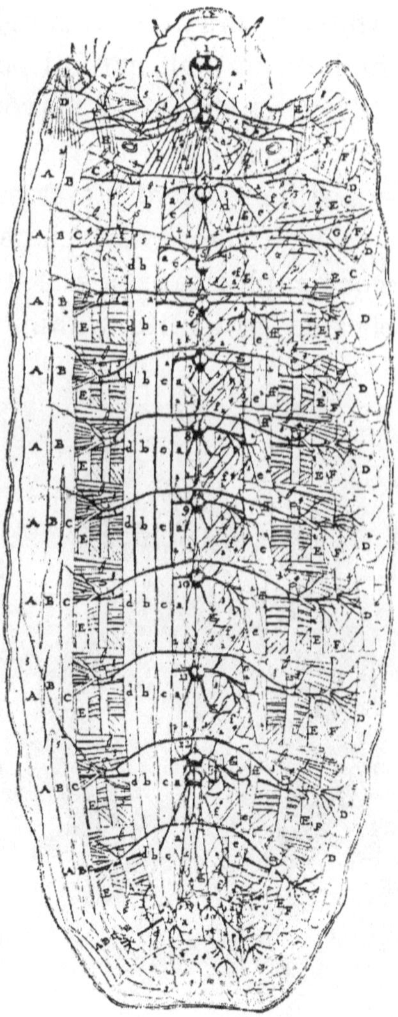

FIG. 25.

FIG. 24.—Muscles of the Larva of the Willow Moth. (From Lyonet's Monograph.)

FIG. 25.—Central Nervous System and Nerves of the Same.

exhibiting the nervous ganglia, the air tubes, and muscles of the head in their natural position.

Fig. 27 shows the nervous system of the head, including the extremely fine nervous masses which are designated the sympathetic nervous system.

The extraordinary character of the drawings in Lyonet's monograph created a sensation. The existence of such complicated structures within the body of an insect was dis-

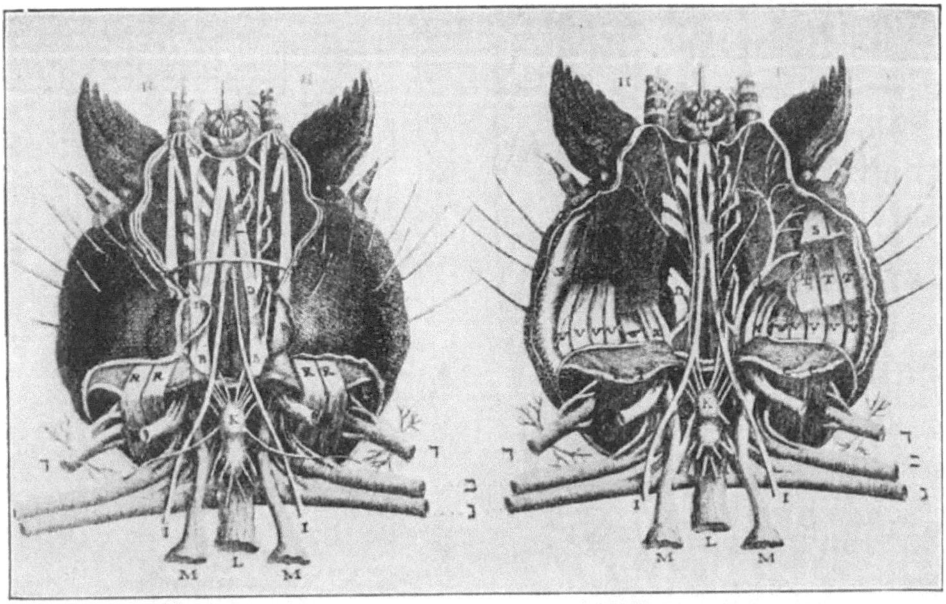

FIG. 26.—Dissection of the Head of the Larva of the Willow Moth.

credited, and, furthermore, some of his critics declared that even if such a fine organization existed, it would be beyond human possibilities to expose the details as shown in his sketches. Accordingly, Lyonet was accused of drawing on his imagination. In order to silence his critics he published in the second edition of his work, in 1752, drawings of his instruments and a description of his methods.

Lyonet intended to work out the anatomy of the chrysalis and the adult form of the same animal. In pursuance of this plan, he made manv dissections and drawings, but, at the age of sixty, on account of the condition of his eyes, he was obliged to stop all close work, and his project remained unfinished. The sketches which he had accumulated were published later, but they fall far short of those illustrating

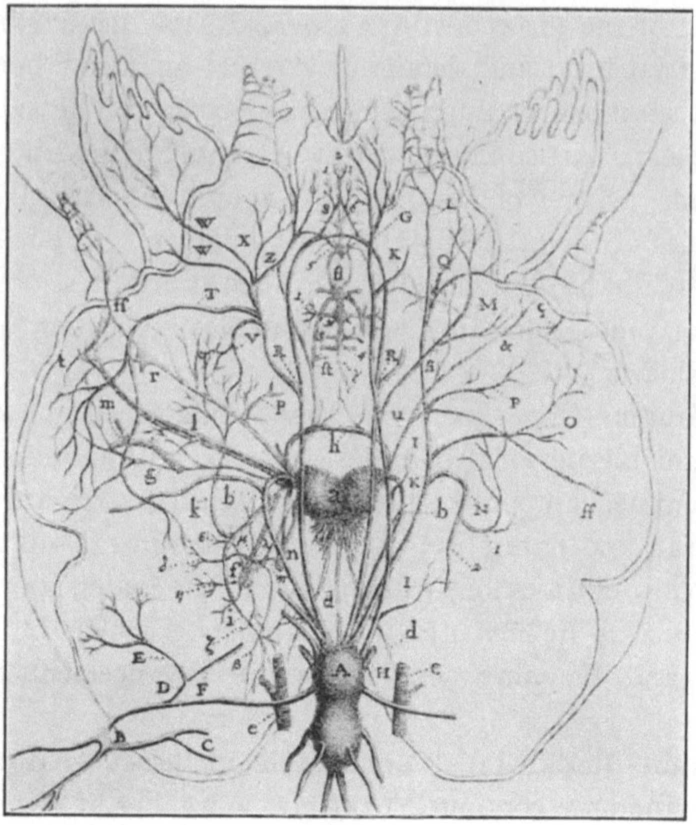

FIG. 27.—The Brain and Head Nerves of the Same Animal.

the *Traité Anatomique.* Lyonet died in 1789, at the age of eighty-one.

Roesel, Réaumur, and De Geer on Insect Life.—We must also take note of the fact that, running parallel with this work on the anatomy of insects, observations and publications had gone forward on form, habits, and metamorphosis of insects, that did more to advance the knowledge of insect life than

Lyonet's researches. Roesel, in Germany, Réaumur, in France, and De Geer, in Sweden, were all distinguished observers in this line. Their works are voluminous and are well illustrated. Those of Réaumur and De Geer took the current French title of *Mémoires pour servir à l'Histoire des Insectes.* The plates with which the collected publications of each of the three men are provided show many sketches of external form and details of external anatomy, but very few illustrations of internal anatomy occur. The sketches of Roesel in particular are worthy of examination at the present time. Some of his masterly figures in color are fine examples of the art of painting in miniature. The name of Roesel (Fig. 28) is connected also with the earliest observations of protoplasm and with a notable publication on the Batrachians.

Réaumur (Fig. 29), who was distinguished for kindly and amiable personal qualities, was also an important man in his influence upon the progress of science. He was both physician and naturalist; he made experiments upon the physiology of digestion, which aided in the understanding of that process; he invented the thermometer which bears his name, and did other services for the advancement of science.

Straus-Dürckheim's Monograph on Insect Anatomy.—Insect anatomy continued to attract a number of observers, but we must go forward into the nineteenth century before we find the subject taking a new direction and merging into its modern phase. The remarkable monograph of Straus-Dürckheim represents the next step in the development of insect anatomy toward the position that it occupies to-day. His aim is clearly indicated in the opening sentence of his preface: "Having been for a long time occupied with the study of articulated animals, I propose to publish a general work upon the comparative anatomy of that branch of the

FIG. 28.—ROESEL VON ROSENHOF, 1705–1759.

animal kingdom." He was working under the influence of Cuvier, who, some years earlier, had founded the science of comparative anatomy and whom he recognized as his great exemplar. His work is dedicated to Cuvier, and is aecom-

FIG. 29.—RÉAUMUR, 1683-1757.

panied by a letter to that great anatomist expressing his thanks for encouragement and assistance.

Straus-Dürckheim (1790–1865) intended that the general considerations should be the chief feature of his monograph, but they failed in this particular because, with the further developments in anatomy, including embryology and the cell-theory, his general discussions regarding the articulated animals became obsolete. The chief value of his work now lies in what he considered its secondary feature, *viz.*, that of the detailed anatomy of the cockchafer, one of the common beetles of Europe. Owing to changed conditions, therefore, it takes rank with the work of Malpighi and Lyonet, as a monograph on a single form. Originally he had intended to publish a series of monographs on the structure of insects typical of the different families, but that upon the cockchafer was the only one completed.

Comparison with the Sketches of Lyonet.—The quality of this work upon the anatomy of the cockchafer was excellent, and in 1824 it was accepted and crowned by the Royal Institute of France. The finely lithographed plates were prepared at the expense of the Institute, and the book was published in 1828 with the following cumbersome title: *Considérations Générales sur l'Anatomie comparée des Animaux Articulés auxquelles on a joint l'Anatomie Descriptive du Melolontha Vulgaris (Hanneton) donnée comme example de l'Organisation des Coléoptères.* The 109 sketches with which the plates are adorned are very beautiful, but one who compares his drawings, figure by figure, with those of Lyonet can not fail to see that those of the latter are more detailed and represent a more careful dissection. One illustration from Straus-Dürckheim will suffice to bring the achievements of the two men into comparison.

Fig. 30 shows his sketch of the anatomy of the central nervous system. He undertakes to show only the main branches of the nerves going to the different segments of the body, while Lyonet brings to view the distribution of the minute terminals to particular muscles. Comparison of other figures—notably that of the dissection of the head—will bring out the same point, *viz.*, that Lyonet was more detailed than Straus-Dürckheim in his explorations of the anatomy of insects, and fully as accurate in drawing what he had seen.

in a different spirit, and is the first serious attempt to make insect anatomy broadly comparative.

Comment.—Such researches as those of Swammerdam, Lyonet, and Straus-Dürckheim represent a phase in the progress of the study of nature. Perhaps their chief value lies in the fact that they embody the idea of critical observation. As examples of faithful, accurate observations the researches helped to bring about that close study which is our only means of getting at basal facts. These men were all enlisted in the crusade against superficial observation. This had to have its beginning, and when we witness it in its early stages, before the researches have become illuminated by great ideas, the prodigious effort involved in the detailed researches may seem to be poorly expended labor. Nevertheless, though the writings of these pioneers have become obsolete, their work was of importance in helping to lift observations upon nature to a higher level.

Dufour.—Léon Dufour extended the work of Straus-Dürckheim by publishing, between 1831 and 1834, researches upon the anatomy and physiology of different families of insects. His aim was to found a general science of insect anatomy. That he was unsuccessful in accomplishing this was owing partly to the absence of embryology and histology from his method of study.

Newport.—The thing most needed now was not greater devotión to details and a willingness to work, but a broadening of the horizon of ideas. This arrived in the Englishman Newport, who was remarkable not only for his skill as a dissector, but for his recognition of the importance of embryology in elucidating the problems of structure. His article "Insecta" in Todd's *Cyclopædia of Anatomy and Physiology*, in 1841, and his papers in the *Philosophical Transactions* of the Royal Society contain this new kind of research.

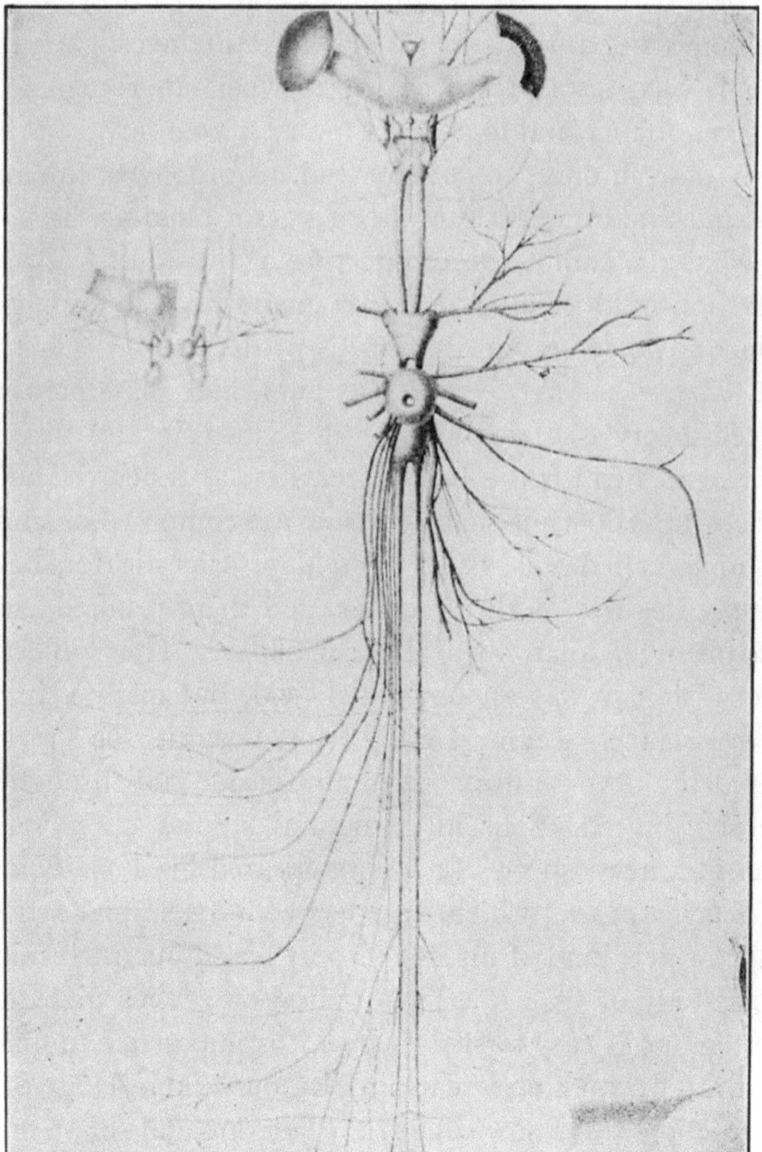

Fig. 30.—Nervous System of the Cockchafer. (From Straus-Dürckheim's Monograph, 1828.)

Von Baer had founded embryology by his great work on the development of animals in 1828, before the investigations of Dufour, but it was reserved for Newport to recognize its great importance and to apply it to insect anatomy. He saw clearly that, in order to comprehend his problems, the anatomist must take into account the process of building the body, as well as the completed architecture of the adult. The introduction of this important idea made his achievement a distinct advance bevond that of his predecessors.

Leydig.—Just as Newport was publishing his conclusions the cell-theory was established (in 1838–39); and this was destined to furnish the basis for a new advance. The influence of the doctrine that all tissues are composed of similar vital units, called cells, was far-reaching. Investigators began to apply the idea in all directions, and there resulted a new department of anatomy, called histology. The subject of insect histology was an unworked field, but manifestly one of importance. Franz Leydig (for portrait see p. 175) entered the new territory with enthusiasm, and through his extensive investigations all structural studies upon insects assumed a new aspect. In 1864 appeared his *Vom Bau des Thierchen Körpers*, which, together with his special articles, created a new kind of insect anatomy based upon the microscopic study of tissues. The application of this method of investigation is easy to see; just as it is impossible to understand the working of a machine without a knowledge of its construction, so a knowledge of the working units of an organ is necessary to comprehend its action. For illustration, it is perfectly evident that we can not understand what is taking place in an organ for receiving sensory impressions without first understanding its mechanism and the nature of the connections between it and the central part of the nervous system. The sensorv organ is on the surface in order more readilv to receive impressions from the outside world. The sensory cells are also modifications of surface cells, and, as a preliminary step to understanding their particular office, we must know the line along which they have become modified to fit them to receive stimulation.

Then, if we attempt to follow in the imagination the way by which the surface stimulations reach the central nervous system and affect it, we must investigate all the connections. It thus appears that we must know the intimate structure of an organ in order to understand its physiology. Leydig supplied this kind of information for many organs of insects. In his investigations we see the foundation of that delicate work upon the microscopic structure of insects which is still going forward.

Summary.—In this brief sketch we have seen that the study of insect anatomy, beginning with that of Malpighi and Swammerdam, was lifted to a plane of greater exactitude by Lyonet and Straus-Dürckheim. It was further broadened by the researches of Dufour, and began to take on its modern aspects, first, through the labors of Newport, who introduced embryology as a feature of investigation, and, finally, through Leydig's step in introducing histology. In the combination of the work of these two observers, the subject for the first time reached its proper position.

The studies of minute structure in the seventeenth and eighteenth centuries were by no means confined to insects; investigations were made upon a number of other forms. Trembley, in the time of Lyonet, produced his noteworthy memoirs upon the small fresh-water hydra *(Mémoires pour servir à l'histoire des polypes d'eau douce,* 1744); the illustrations for which, as already stated, were prepared by Lyonet. The structure of snails and other mollusks, of tadpoles, frogs, and other batrachia, was also investigated. We have seen that Swammerdam, in the seventeenth century, had begun observations upon the anatomy of tadpoles, frogs, and snails, and also upon the minute crustacea commonly called water-fleas, which are just large enough to be distinguished by the unaided eye. We should remember also that in the same period the microscopic structure of plants began to be investigated, notably by Grew, Malpighi, and Leeuwenhoek (see Chapter IV).

In addition to those essays into minute anatomy, in which scalpel and scissors were employed, an endeavor of more subtle difficulty made its appeal; there were forms of animal life of still smaller size and simpler organization that began to engage the attention of microscopists. A brief account of the discovery and subsequent observation of these microscopic animalcula will now occupy our attention.

The Discovery of the Simplest Animals and the Progress of Observations upon Them

These single-celled animals, since 1845 called protozoa, have become of unusual interest to biologists, because in them the processes of life are reduced to their simplest expression. The vital activities taking place in the bodies of higher animals are too complicated and too intricately mixed to admit of clear analysis, and, long ago, physiologists learned that the quest for explanations of living activities lay along the line of investigating them in their most rudimentary expression. The practical recognition of this is seen in our recent text-books upon human physiology, which commonly begin with discussions of the life of these simplest organisms. That greatest of all text-books on general physiology, written by Max Verworn, is devoted largely to experimental studies upon these simple organisms as containing the key to the similar activities (carried on in a higher degree) in higher animals. This group of animals is so important as a field of experimental observation that a brief account of their discovery and the progress of knowledge in reference to them will be in place in this chapter.

Discovery of the Protozoa.—Leeuwenhoek left so little unnoticed in the microscopic world that we are prepared to find that he made the first recorded observations upon these animalcula. His earliest observations were communicated by letter to the Royal Society of London, and were published in their *Transactions* in 1677. It is very interesting to read his descriptions expressed in the archaic language of the time. The following quotation from a Dutch letter turned into English will suffice to give the flavor of his writing:

"In the year 1675 I discovered living creatures in rain-water which had stood but four days in a new earthen pot, glazed blew within. This invited me to view the water with great attention, especially those little animals appearing to me ten thousand times less than those represented by Mons. Swammerdam, and by him called water-fleas or water-lice, which may be perceived in the water with the naked eye. The first sorte by me discovered in the said water, I divers times observed to consist of five, six, seven or eight clear globules, without being able to discover any film that held them together or contained them. When these *animalcula*, or living atoms, did move they put forth two little horns, continually moving themselves; the place between these two horns was flat, though the rest of the body was roundish, sharpening a little towards the end, where they had a tayle, near four times the length of the whole body, of the thickness (by my microscope) of a spider's web; at the end of which appeared a globule. of the bigness of one of those which made up the body; which tayle I could not perceive even in very clear water to be mov'd by them. These little creatures, if they chanced to light upon the least filament or string, or other such particle, of which there are many in the water, especially after it has stood some days, they stook entangled therein, extending their body in a long round, and striving to dis-entangle their tayle; whereby it came to pass, that their whole body lept back towards the globule of the tayle, which then rolled together serpent-like, and after the manner of copper or iron wire, that having been wound around a stick, and unwound again, retains those windings and turnings," etc.*

Any one who has examined under the microscope the well-known bell-animalcule will recognize in this first description of it, the stalk, and its form after contraction under the designation of a 'tayle which retains those windings and turnings.'

There are many other descriptions, but the one given is typical of the others. He found the little animals in water, in infusions of pepper, and other vegetable substances, and on that account they came soon to be designated infusoria. His observations were not at first accompanied by sketches, but in 1711 he sent some drawings with further descriptions.

O. Fr. Müller.—These animalcula became favorite objects of microscopic study. Descriptions began to accumulate and drawings to be made until it became evident that there were many different kinds. It was, however, more than one hundred years after their discovery by Leeuwenhoek that the first standard work devoted exclusively to these animalcula was published. This treatise by O. Fr. Müller was published in 1786 under the title of *Animalcula Infusoria.* The circumstance that this volume of quarto size had 367 pages of description with 50 plates of sketches will give some indication of the number of protozoa known at that time.

Ehrenberg.—Observations in this domain kept accumulating, but the next publication necessary to mention is that of Ehrenberg (1795–1876). This scientific traveler and eminent observer was the author of several works. He was

* Kent's Manual of the Infusoria, Vol. I, p. 3. Quotation from the *Philosophical Transactions* for the year 1677.

one of the early observers of nerve fibres and of many other structures of the animal frame. His book of the protozoa is a beautifully illustrated monograph consisting of 532 pages of letterpress and 69 plates of folio size. It was published in 1836 under the German title of *Die Infusionsthierchen als Vollkommene Organismen*, or "The Infusoria as Perfect Organisms." The animalcula which he so faithfully represented in his sketches have the habit, when feeding, of taking into the body collections of food-particles, aggregated into spherical globules or food vacuoles. These are distinctly separated, and slowly circulate around the single-celled body while they are undergoing digestion. In a fully fed animal these food-vacuoles occupy different positions, and are enclosed in globular spaces in the protoplasm, an adjustment that gave Ehrenberg the notion that the animals possessed many stomachs. Accordingly he gave to them the name "Polygastrica," and assigned to them a much higher grade of organization than they really possess. These conclusions, based on the general arrangement of food globules, seem very curious to us to-day. His publication was almost simultaneous with the announcement of the cell-theory (1838–39), the acceptance of which was destined to overthrow his conception of the protozoa, and to make it clear that tissues and organs can belong only to multicellular organisms.

Ehrenberg (Fig. 31) was a man of great scientific attainments, and notwithstanding the grotesqueness of some of his conclusions, was held in high esteem as a scientific investigator. His observations were accurate, and the beautiful figures with which his work on the protozoa is embellished were executed with such fidelity regarding fine points of microscopic detail that they are of value to-day.

Dujardin, whom we shall soon come to know as the discoverer of protoplasm, successfully combated the conclusions of Ehrenberg regarding the organization of the protozoa.

For a time the great German scientist tried to maintain his point, that the infusoria have many stomachs, but this was completely swept away, and finally the contention of Von Siebold was adopted to the effect that these animals are each composed of a single cell.

In 1845 Stein is engrossed in proposing names for the suborders of infusoria based upon the distribution of cilia

FIG. 31.—EHRENBERG, 1795–1876.

upon their bodies. This simple method of classification as well as the names introduced by Stein, is still in use.

From Stein to Bütschli, one of the present authorities on the group, there were many workers, but with the studies of Bütschli on protozoa we enter the modern epoch.

The importance of these animals in affording a field for experimentation on the simplest expressions of life has already been indicated. Many interesting problems have arisen in connection with recent studies of them. The group embraces the very simplest manifestations of animal life, and the experiments upon the different forms light the way for studies of the vital activities of the higher animals. Some of the protozoa are disease-producing; as the microbe of malaria, of the sleeping sickness, etc., while, as is well known, most diseases that have been traced to specific germs are caused by plants—the bacteria. Many experiments of Maupas, Caulkins and others have a bearing upon the dis cussions regarding the immortality of the protozoa, an idea which was at one time a feature of Weissmann's theory of heredity. Binet and others have discussed the evidences of psychic life in these micro-organisms, and the daily activity of a protozoan became the field for observation and record in an American laboratory of psychology. The extensive studies of Jennings on the nature of their responses to stimulations form a basis for some of the discussions on animal behavior.

CHAPTER VI

LINNÆUS AND SCIENTIFIC NATURAL HISTORY

We turn now from the purely anatomical side to consider the parallel development of the classification of animals and of plants. Descriptive natural history reached a very low level in the early Christian centuries, and remained there throughout the Middle Ages. The return to the writings of Aristotle was the first influence tending to lift it to the position from which it had fallen. After the decline of ancient civilization there was a period in which the writers of classical antiquity were not read. Not only were the writings of the ancient philosophers neglected, but so also were those of the literary men as well, the poets, the story-tellers, and the historians. As related in Chapter I, there were no observations of animated nature, and the growing tendency of the educated classes to envelop themselves in metaphysical speculations was a feature of intellectual life.

The Physiologus or Sacred Natural History.—During this period of crude fancy, with a fog of mysticism obscuring all phenomena of nature, there existed a peculiar kind of natural history that was produced under theological influence. The manuscripts in which this sacred natural history was embodied exist in various forms and in about a dozen languages of Eastern and Western Europe. The writings are known under the general title of the Physiologus, or the Bestiarius. This served for nearly a thousand years as the principal source of thought regarding natural history. It contains accounts of animals mentioned in the Bible and others of a purely mythical character. These are made to be symbolical of religious beliefs, and are often accompanied by quotations of texts and by moral reflections. The phœnix rising from its ashes typifies the resurrection of Christ. In reference to young lions, the *Physiologus* says: "The lioness giveth birth to cubs which remain three days without life. Then cometh the lion, breatheth upon them, and bringeth them to life. Thus it is that Jesus Christ during three days was deprived of life, but God the Father raised him gloriously." (Quoted from White, p. 35.) Besides forty or fifty common animals, the unicorn and the dragon of the Scriptures, and the fabled basilisk and phœnix of secular writings are described, and morals are drawn from the stories about them. Some of the accounts of animals, as the lion, the panther, the serpent, the weasel, etc., etc., are so curious that, if space permitted, it would be interesting to quote them; but that would keep us too long from following the rise of scientific natural history from this basis.

For a long time the religious character of the contemplations of nature was emphasized and the prevalence of theological influence in natural history is shown in various titles, as Lesser's *Theology of Insects*, Swammerdam's *Biblia Naturæ*, Spallanzani's *Tracts*, etc.

The zoölogy of the *Physiologus* was of a much lower grade than any we know about among the ancients, and it is a curious fact that progress was made by returning to the natural history of fifteen centuries in the past. The translation of Aristotle's writings upon animals, and the disposition to read them, mark this advance. When, in the Middle Ages, the boundaries of interest began to be extended, it came like an entirely new discovery, to find in the writings of the ancients a storehouse of philosophic thought and a higher grade of learning than that of the period. The translation and recopying of the writers of classical antiquity was, therefore, an important step in the revival of learning. These writings were so much above the thought of the time that the belief was naturally created that the ancients had digested all learning, and they were pointed to as unfailing authorities in matters of science.

The Return to the Science of the Ancients.—The return to Aristotle was wholesome, and under its influence men turned their attention once more to real animals. Comments upon Aristotle began to be made, and in course of time independent treatises upon animals began to appear. One of the first to modify Aristotle to any purpose was Edward Wotton, the English physician, who published in 1552 a book on the distinguishing characteristics of animals (*De Differentiis Animalium*). This was a complete treatise on the zoölogy of the period, including an account of the different races of mankind. It was beautifully printed in Paris, and was dedicated to Edward VI. Although embracing ten books, it was by no means so ponderous as were some of the treatises that followed it. The work was based upon Aristotle, but the author introduced new matter, and also added the group of zoöphytes, or plant-like animals of the sea.

Gesner.—The next to reach a distinctly higher plane was Conrad Gesner (1516–1565), the Swiss, who was a contemporary of Vesalius. He was a practising physician who, in 1553, was made professor of natural history in Zurich. A man of extraordinary talent and learning, he turned out an astonishing quantity of work. Besides accomplishing much in scientific lines, he translated from Greek, Arabic, and Hebrew, and published in twenty volumes a universal catalogue of all works known in Latin, Greek, and Hebrew, either printed or in manuscript form. In the domain of natural history he began to look critically at animals with a view to describing them, and to collect with zealous care new observations upon their habits. His great work on natural history (*Historia Animalium*) began to appear in 1551, when he was thirty-five years of age, and four of the five volumes were published by 1556. The fifth volume was not published until 1587, twenty-two years after his death. The complete work consists of about "4,500 folio pages," profusely illustrated with good figures. The edition which the writer has before him—that of 1585–1604—embraces 3,200 pages of text and 953 figures.

Brooks says: "One of Gesner's greatest services to natural science is the introduction of good illustrations, which he gives his reader by hundreds." He was so exacting about the quality of his illustrations that his critical supervision of the work of artists and engravers had its influence upon contemporary art. Some of the best woodcuts of the period are found in his work. His friend Albrecht Dürer supplied one of the originals—the drawing of the rhinoceros—and it is interesting to note that it is by no means the best, a circumstance which indicates how effectively Gesner held his engraver and draughtsman up to fine work. He was also careful to mold his writing into graceful form, and this, combined with the illustrations, "made science attractive without sacrificing its dignity, and thus became a great educational influence."

In preparing his work he sifted the writings of about two hundred and fifty authors, and while his book is largely a compilation, it is enriched with many observations of his own. His descriptions are verbose, but discriminating in separating facts and observations from fables and speculations. He could not entirely escape from old traditions. There are retained in his book pictures of the sea-serpent, the mermaids, and a few other fanciful and grotesque sketches, but for the most part the drawings are made from the natural objects. The descriptions are in several parts of his work alphabetically arranged, for convenience of reference, and thus animals that were closely related are often widely separated.

Gesner (Fig. 32) sacrificed his life to professional zeal during the prevalence of the plague in Zurich in 1564. Having greatly overworked in the care of the sick, he was seized with the disease, and died at the age of forty-nine.

Considered from the standpoint of descriptions and illustrations, Gesner's *Historia Animalium* remained for a long

FIG. 32.—GESNER 1516–1565.

time the best work in zoölogy. He was the best zoölogist between Aristotle and John Ray, the immediate predecessor of Linnæus.

Jonston and Aldrovandi.—At about the same period as Gesner's work there appeared two other voluminous publications, which are well known—those of Jonston, the Scot

(*Historia Animalium*, 1549–1553), and Aldrovandi, the Italian (*Opera*, 1599–1606). The former consisted of four folio volumes, and the latter of thirteen, of ponderous size, to which was added a fourteenth on plants. Jonston's works were translated, and were better known in England than those of Gesner and Aldrovandi. The wood-engravings in Aldrovandi's volume are coarser than those of Gesner, and are by no means so lifelike. In the Institute at Bologna are preserved twenty volumes of figures of animals in color, which were the originals from which the engravings were made. These are said to be much superior to the reproductions. The encyclopædic nature of the writings of Gesner, Aldrovandi, and Jonston has given rise to the convenient and expressive title of the encyclopædists.

Ray.—John Ray, the forerunner of Linnæus, built upon the foundations of Gesner and others, and raised the natural-history edifice a tier higher. He greatly reduced the bulk of publications on natural history, sifting from Gesner and Aldrovandi their irrelevancies, and thereby giving a more modern tone to scientific writings. He was the son of a blacksmith, and was born in southern England in 1628. The original form of the family name was Wray. He was graduated at the University of Cambridge, and became a fellow of Trinity College. Here he formed a friendship with Francis Willughby, a young man of wealth whose tastes for natural history were like his own. This association proved a happy one for both parties. Ray had taken orders in the Church of England, and held his university position as a cleric; but, from conscientious scruples, he resigned his fellowship in 1662. Thereafter he received financial assistance from Willughby, and the two men traveled extensively in Great Britain and on the Continent, with the view of investigating the natural history of the places that they visited. On these excursions Willughby gave particular attention to animals and Ray to plants. Of Ray's several publications in botany, his *Historia Plantarum* in three volumes (1686-1704) is the most extensive. In another work, as early as 1682, he had proposed a new classification of plants, which

FIG. 33.—JOHN RAY, 1628-1705.

in the next century was adopted by Jussieu, and which gives Ray a place in the history of botany.

Willughby died in 1662, at the age of thirty-eight, leaving an annuity to Ray, and charging him with the education of his two sons, and the editing of his manuscripts. Ray performed these duties as a faithful friend and in a generous spirit. He edited and published Willughby's book on birds (1678) and fishes (1686) with important additions of his own, for which he sought no credit.

After completing his tasks as the literary executor of Willughby, he returned in 1678 to his birthplace and continued his studies in natural history. In 1691 he published "The Wisdom of God manifested in the Works of the Creation," which was often reprinted, and became the forerunner of the works on natural theology like Paley's, etc. This was an amplification of ideas he had embodied in a sermon thirty-one years earlier, and which at that time attracted much notice. He now devoted himself largely to the study of animals, and in 1693 published a work on the quadrupeds and serpents, a work which gave him high rank in the history of the classification of animals. He died in 1705, but he had accomplished much good work, and was not forgotten. In 1844 there was founded, in London, in his memory, the Ray Society for the publication of rare books on botany and zoölogy.

Ray's Idea of Species.—One of the features of Ray's work, in the light of subsequent development, is of special interest, and that is his limiting of species. He was the first to introduce into natural history an exact conception of species. Before his time the word had been used in an indefinite sense to embrace groups of greater or less extent, but Ray applied it to individuals derived from similar parents, thus making the term species stand for a particular kind of animal or plant. He noted some variations among species, and did not assign to them that unvarying and constant character ascribed to them by Linnæus and his followers. Ray also made use of anatomy as the foundation for zoölogical classification, and introduced great precision and clearness into his definitions of groups of animals and plants. In the particulars indicated above he represents a great advance beyond any of his precursors, and marks the parting of the ways between mediæval and modern natural history.

In Germany Klein (1685–1759) elaborated a system of classification embracing the entire animal kingdom. His studies were numerous, and his system would have been of much wider influence in molding natural history had it not been overshadowed by that of Linnæus.

Linnæus or Linné.—The service of Linnæus to natural history was unique. The large number of specimens of animals and plants, ever increasing through the collections of travelers and naturalists, were in a confused state, and there was great ambiguity arising from the lack of a methodical way of arranging and naming them. They were known by verbose descriptions and local names. No scheme had as yet been devised for securing uniformity in applying names to them. The same animal and plant had different names in the different sections of a country, and often different plants and animals had the same name. In different countries, also, their names were greatly diversified. What was especially needed was some great organizing mind to catalogue the animals and plants in a systematic way, and to give to natural science a common language. Linnæus possessed this methodizing mind and supplied the need. While he did little to deepen the knowledge of the organization of animal and plant life, he did much to extend the number of known forms; he simplified the problem of cataloguing them, and he invented a simple method of naming them which was adopted throughout the world. By a happy stroke he gave to biology a new language that remains in use to-day. The tremendous influence of this may be realized when we remember that naturalists everywhere use identical names for the same animals and plants. The residents of Japan, of Italy, of

Spain, of all the world, in fact, as was just said, employ the same Latin names in classifying organic forms.

He also inspired many students with a love for natural history and gave an impulse to the advance of that science which was long felt. We can not gainsay that a higher class of service has been rendered by those of philosophic mind devoted to the pursuit of comparative anatomy, but the step of Linnæus was a necessary one, and aided greatly in the progress of natural history. Without this step the discoveries and observations of others would not have been so readily understood, and had it not been for his organizing force all natural science would have been held back for want of a common language. A close scrutiny of the practice among naturalists in the time of Linnæus shows that he did not actually invent the binomial nomenclature, but by adopting the suggestions of others he elaborated the system of classification and brought the new language into common use.

Personal History.—Leaving for the present the system of Linnæus, we shall give attention to the personal history of the man. The great Swedish naturalist was born in Rashult in 1707. His father was the pastor of the village, and intended his eldest son, Carl, for the same high calling. The original family name was Ignomarsen, but it had been changed to Lindelius, from a tall linden-tree growing in that part of the country. In 1761 a patent of nobility was granted by the crown to Linnæus, and thereafter he was styled Carl von Linné.

His father's resources were very limited, but he managed to send his son to school, though it must be confessed that young Linnæus showed little liking for the ordinary branches of instruction. His time was spent in collecting natural-history specimens, and his mind was engaged in thinking about them. The reports of his low scholarship and the statement of one of his teachers that he showed no aptitude for learning were so disappointing to his father that, in 1726, he prepared to apprentice Carl to a shoemaker, but was prevented from doing so through the encouragement of a doctor who, being able to appreciate the quality of mind possessed by the young Linnæus, advised allowing him to study medicine instead of preparing for theology.

Accordingly, with a sum amounting to about $40, all his father could spare, he set off for the University of Lund, to pursue the study of medicine. He soon transferred to the University of Upsala, where the advantages were greater. His poverty placed him under the greatest straits for the necessities of life, and he enjoyed no luxuries. While in the university he mended his shoes, and the shoes which were given to him by some of his companions, with paper and birch-bark, to keep his feet from the damp earth. But his means did not permit of his taking his degree at Upsala, and it was not until eight years later, in 1735, that he received his degree in Holland.

At Upsala he was relieved from his extreme poverty by obtaining an assistant's position, and so great was his knowledge of plants that he was delegated to read the lectures of the aged professor of botany, Rudbeck.

In 1732 he was chosen by the Royal Society of Upsala to visit Lapland as a collector and observer, and left the university without his degree. On returning to Upsala, his lack of funds made itself again painfully felt, and he undertook to support himself by giving public lectures on botany, chemistry, and mineralogy. He secured hearers, but the continuance of his lectures was prevented by one of his rivals on the ground that Linnæus had no degree, and was therefore legally disqualified from taking pay for instruction. Presently he became tutor and traveling companion of a wealthy baron, the governor of the province of Dalecarlia, but this employment was temporary.

Helped by His Fiancée.—His friends advised him to secure his medical degree and settle as a practitioner. Although he lacked the necessary funds, one circumstance contributed to bring about this end: he had formed an attachment for the daughter of a wealthy physician, named Moré or Moræus, and on applying for her hand in marriage, her father made it a condition of his consent that Linnæus should take his medical degree and establish himself in the practice of medicine. The young lady, who was thrifty as well as handsome, offered her savings, amounting to one hundred dollars (Swedish), to her lover. He succeeded in adding to this sum by his own exertions, and with thirty-six Swedish ducats set off for Holland to qualify for his degree. He had practically met the requirements for the medical degree by his previous studies, and after a month's residence at the University of Hardewyk, his thesis was accepted and he was granted the degree in June, 1735, in the twenty-eighth year of his age.

Instead of returning at once to Sweden, he went to Leyden, and made the acquaintance of several well-known scientific men. He continued his botanical studies with great energy, and now began to reap the benefits of his earlier devotion to natural history. His heart-breaking and harassing struggles were now over.

The Systema Naturæ.—He had in his possession the manuscript of his *Systema Naturæ*, and with the encouragement of his new friends it was published in the same year. The first edition (1735) of that notable work, which was afterward to bring him so much fame, consisted of twelve printed folio pages. It was merely an outline of the arrangements of plants, animals, and minerals in a methodical catalogue. This work passed through twelve editions during his lifetime, the last one appearing in 1768. After the first edition, the books were printed in octavo form, and in the later editions were greatly enlarged. A copy of the first edition was sent to Boerhaave, the most distinguished professor in the University of Leyden, and secured for Linnæus an interview with that distinguished physician, who treated him with consideration and encouraged him in his work. Boerhaave was already old, and had not long to live; and when Linnæus was about to leave Holland in 1738, he admitted him to his sick-chamber and bade him a most affectionate adieu, and encouraged him to further work by most kindly and appreciative expressions.

Through the influence of Boerhaave, Linnæus became the medical attendant of Cliffort, the burgomaster at Amsterdam, who had a large botanic garden. Cliffort, being desirous of extending his collections, sent Linnæus to England, where he met Sir Hans Sloane and other eminent scientific men of Great Britain. After a short period he returned to Holland, and in 1737 brought out the *Genera Plantarum*, a very original work, containing an analysis of all the genera of plants. He had previously published, besides the *Systema Naturæ*, his *Fundamenta Botanica*, 1735, and *Bibliotheca Botanica*, 1736, and these works served to spread his fame as a botanist throughout Europe.

His Wide Recognition.—An illustration of his wide recognition is afforded by an anecdote of his first visit to Paris in 1738. "On his arrival he went first to the Garden of Plants, where Bernard de Jussieu was describing some exotics in Latin. He entered without opportunity to introduce himself. There was one plant which the demonstrator had not yet determined, and which seemed to puzzle him. The Swede looked on in silence, but observing the hesitation of the learned professor, cried out '*Hæc planta faciem Americanam habet.*' 'It has the appearance of an American plant.' Jussieu, surprised, turned about quickly and exclaimed 'You are Linnæus.' 'I am, sir,' was the reply. The lecture was stopped, and Bernard gave the learned stranger an affectionate welcome."

Return to Sweden.—After an absence of three and one-half years, Linnæus returned to his native country in 1738, and soon after was married to the young woman who had assisted him and had waited for him so loyally. He settled in Stockholm and began the practice of medicine. In the period of his absence he had accomplished much: visited Holland, England, and France, formed the acquaintance of many eminent naturalists, obtained his medical degree, published numerous works on botany, and extended his fame over all Europe. In Stockholm, however, he was for a time neglected, and he would have left his native country in disgust had it not been for the dissuasion of his wife.

Professor in Upsala.—In 1741 he was elected professor of anatomy in the University of Upsala, but by a happy stroke was able to exchange that position for the professorship of botany, materia medica, and natural history that had fallen to his former rival, Rosen. Linnæus was now in his proper element; he had opportunity to lecture on those subjects to which he had been devotedly attached all his life, and he entered upon the work with enthusiasm.

He attracted numerous students by the power of his personal qualities and the excellence of his lectures. He became the most popular professor in the University of Upsala, and, owing to his drawing power, the attendance at the university was greatly increased. In 1749 he had 140 students devoted to studies in natural history. The number of students at the university had been about 500; "whilst he occupied the chair of botany there it rose to 1,500." A part of this increase was due to other causes, but Linnæus was the greatest single drawing force in the university. He was an eloquent as well as an enthusiastic lecturer, and he aroused great in terest among his students, and he gave an astonishing impulse to the study of natural history in general, and to botany in particular. Thus Linnæus, after having passed through great privations in his earlier years, found himself, at the age of thirty-four, established in a position which brought him recognition, honor, and large emolument.

In May, 1907, the University of Upsala celebrated the two hundredth anniversary of his birth with appropriate ceremonies.

FIG. 34.—LINNÆUS AT SIXTY, 1707–1778.

Delegations of scientific men from all over the world were in attendance to do honor to the memory of the great founder of biological nomenclature.

Personal Appearance.—The portrait of Linnæus at the age of sixty is shown in Fig. 34. He was described as of "medium height, with large limbs, brown, piercing eyes, and acute vision." His hair in early youth was nearly white, and changed in his manhood to brown, and became gray with the advance of age. Although quick-tempered, he was naturally of a kindly disposition, and secured the affection of his students, with whom he associated and worked in the most informal way. His love of approbation was very marked, and he was so much praised that his desire for fame became his dominant passion. The criticism to which his work was subjected from time to time accordingly threw him into fits of despondency and rage.

His Influence upon Natural History.—However much we may admire the industry and force of Linnæus, we must admit that he gave to natural history a one-sided development, in which the more essential parts of the science received scant recognition. His students, like their master, were mainly collectors and classifiers. "In their zeal for naming and classifying, the higher goal of investigation, knowledge of the nature of animals and plants, was lost sight of and the interest in anatomy, physiology, and embryology lagged."

R. Hertwig says of him: "For while he in his *Systema Naturæ* treated of an extraordinarily larger number of animals than any earlier naturalist, he brought about no deepening of our knowledge. The manner in which he divided the animal kingdom, in comparison with the Aristotelian system, is to be called rather a retrogression than an advance. Linnæus divided the animal kingdom into six classes—Mammalia, Aves, Amphibia, Pisces, Insecta, Vermes. The first four classes correspond to Aristotle's four groups of animals with blood. In the division of the invertebrated animals into Insecta and Vermes Linnæus stands undoubtedly behind

Aristotle, who attempted, and in part indeed successfully, to set up a larger number of groups.

"But in his successors even more than in Linnæus himself we see the damage wrought by the purely systematic method of consideration. The diagnoses of Linnæus were for the most part models, which, *mutatis mutandis*, could be employed for new species with little trouble. There was needed only some exchanging of adjectives to express the differences. With the hundreds of thousands of different species of animals, there was no lack of material, and so the arena was opened for that spiritless zoölogy of species-making, which in the first half of the nineteenth century brought zoölogy into such discredit. Zoölogy would have been in danger of growing into a Tower of Babel of species-description if a counterpoise had not been created in the strengthening of the physiologico-anatomical method of consideration."

His Especial Service.—Nevertheless, the work of Linnæus made a lasting impression upon natural history, and we shall do well to get clearly in mind the nature of his particular service. In the first place, he brought into use the method of naming animals and plants which is employed to-day. In his *Systema Naturæ* and in other publications he employed a means of naming every natural production in two words, and it is therefore called the binomial nomenclature. An illustration will make this clearer. Those animals which had close resemblance, like the lion, tiger, leopard, the lynx, and the cat, he united under the common generic name of *Felis*, and gave to each a particular trivial name, or specific name. Thus the name of the lion became *Felis leo*, of the tiger *Felis tigris*, of the leopard *Felis pardus*, of the cat *Felis catus*; and to these the modern zoölogists have added, making the Canada lynx *Felis Canadensis*, the domestic cat *Felis domesticata*, etc. In a similar way, the dog-like animals were united into a genus designated *Canis*, and the particular kinds or species became *Canis lupus*, the wolf, *Canis vulpes*, the fox, *Canis familiaris*, the common dog. This simple method took the place of the varying names applied to the same animal in different countries and local names in the same country. It recognized at once their generic likeness and their specific individuality.

All animals, plants, and minerals were named according to this method. Thus there were introduced into nomenclature two groups, the genus and the species. The name of the genus was a noun, and that of the species an adjective agreeing with it. In the choice of these names Linnæus sought to express some distinguishing feature that would be suggestive of the particular animal, plant, or mineral. The trivial, or specific, names were first employed by Linnæus in 1749, and were introduced into his *Species Plantarum* in 1753, and into the tenth edition of his *Systema Naturæ* in 1758.

We recognize Linnæus as the founder of nomenclature in natural history, and by the common consent of naturalists the date 1758 has come to be accepted as the starting-point for determining the generic and specific names of animals. The much vexed question of priority of names for animals is settled by going back to the tenth edition of his *Systema Naturæ*, while the botanists have adopted his *Species Plantarum*, 1753, as their base-line for names. As to his larger divisions of animals and plants, he recognized classes and orders. Then came genera and species. Linnæus did not use the term family in his formulæ; this convenient designation was first used and introduced in 1780 by Batch.

The *Systema Naturæ* is not a treatise on the organization of animals and plants; it is rather a catalogue of the productions of nature methodically arranged. His aim in fact was not to give full descriptions, but to make a methodical arrangement.

To do justice, however, to the discernment of Linnæus, it should be added that he was fully aware of the artificial nature of his classification. As Kerner has said: "It is not the fault of this accomplished and renowned naturalist if a greater importance were attached to his system than he himself ever intended. Linnæus never regarded his twenty-four classes as real and natural divisions of the vegetable kingdom, and specifically says so; it was constructed for convenience of reference and identification of species. A real natural system, founded on the true affinities of plants as indicated by the structural characters, he regarded as the highest aim of botanical endeavor. He never completed a natural system, leaving only a fragment (published in 1738)."

Terseness of Descriptions.—His descriptions were marked by extreme brevity, but by great clearness. This is a second feature of his work. In giving the diagnosis of a form he was very terse. He did not employ fullv formed sentences containing a verb, but words concisely put together so as to bring out the chief things he wished to emphasize. As an illustration of this, we may take his characterization of the forest rose, "*Rosa sylvestris vulgaris, flore odorata incarnato.*" The common rose of the forest with a flesh-colored, sweet-smelling flower. In thus fixing the attention upon essential points he got rid of verbiage, a step that was of very great importance.

His Idea of Species.—A third feature of his work was that of emphasizing the idea of species. In this he built upon the work of Ray. We have already seen that Ray was the first to define species and to bring the conception into natural historv. Ray had spoken of the variability of species, but Linnæus, in his earlier publications, declared that they were constant and invariable. His conception of a species was that of individuals born from similar parents. It was assumed that at the original stocking of the earth, one pair of each kind of animals was created, and that existing species were the direct descendants without change of form or habit from the original pair. As to their number, he said: "*Species tot sunt, quot formæ ab initio creatæ sunt*"—there are just so many species as there were forms created in the beginning; and his oft-quoted remark, "*Nulla species nova*," indicates in terse language his position as to the formation of new species. Linnæus took up this idea as expressing the current thought, without analysis of what was involved in it. He readily might have seen that if there were but a single pair of each kind, some of them must have been sacrificed to the hunger of the carnivorous kinds; but, better than making any theories, he might have looked for evidence in nature as to the fixity of species.

While Linnæus first pronounced upon the fixity of species, it is interesting to note that his extended observations upon nature led him to see that variation among animals and plants is common and extensive, and accordingly in the later editions of his *Systema Naturæ* we find him receding from the position that species are fixed and constant. Nevertheless, it was owing to his influence, more than to that of any other writer of the period, that the dogma of fixity of species was established. His great contemporary Buffon looked upon species as not having a fixed reality in nature, but as being figments of the imagination; and we shall see in a later section of this book how the idea of Linnæus in reference to the fixity of species gave way to accumulating evidence on the matter.

Summary.—The chief services of Linnæus to natural science consisted of these three things: bringing into current use the binomial nomenclature, the introduction of terse formulæ for description, and fixing attention upon species. The first two were necessary steps; they introduced clearness and order into the management of the immense number of details, and they made it possible for the observations and discoveries of others to be understood and to take their place in the great system of which he was the originator. The effect of the last step was to direct the attention of naturalists to species, and thereby to pave the way for the coming consideration of their origin, a consideration which became such a burning question in the last half of the nineteenth century.

Reform of the Linnæan System

Necessity of Reform.—As indicated above, the classification established by Linnæus had grave defects; it was not founded on a knowledge of the comparative structure of animals and plants, but in many instances upon superficial features that were not distinctive in determining their position and relationships. His system was essentially an artificial one, a convenient key for finding the names of animals and plants, but doing violence to the natural arrangement of those organisms. An illustration of this is seen in his classification of plants into classes, mainly on the basis of the number of stamens in the flower, and into orders according to the number of pistils. Moreover, the true object of investigation was obscured by the Linnæan system. The chief aim of biological study being to extend our knowledge of the structure, development, and physiology of animals and plants as a means of understanding more about their life, the arrangement of animals and plants into groups should be the outcome of such studies rather than an end in itself.

It was necessary to follow different methods to bring natural history back into the line of true progress. The first modification of importance to the Linnæan system was that of Cuvier, who proposed a grouping of animals based upon a knowledge of their comparative anatomy. He declared that animals exhibit four types of organization, and his types were substituted for the primary groups of Linnæus.

The Scale of Being.—In order to understand the bearing of Cuvier's conclusions we must take note of certain views regarding the animal kingdom that were generally accepted at the time of his writing. Between Linnæus and Cuvier there had emerged the idea that all animals, from the lowest to the highest, form a graduated series. This grouping of animals into a linear arrangement was called exposing the Scale of Being, or the Scale of Nature (*Scala Naturæ*). Buffon, Lamarck, and Bonnet were among the chief exponents of this idea.

That Lamarck's connection with it was temporary has been generally overlooked. It is the usual statement in the histories of natural science, as in the *Encyclopædia Britannica*, in the History of Carus, and in Thomson's *Science of Life*, that the idea of the scale of nature found its fullest expression in Lamarck. Thomson says: "His classification (1801–1812) represents the climax of the attempt to arrange the groups of animals in linear order from lower to higher, in what was called a *scala naturæ*" (p. 14). Even so careful a writer as Richard Hertwig has expressed the matter in a similar form. Now, while Lamarck at first adopted a linear classification, it is only a partial reading of his works that will support the conclusion that he held to it. In his *Système des Animaux sans Vertèbres*, published in 1801, he arranged animals in this way; but to do credit to his discernment, it should be observed that he was the first to employ a genealogical tree and to break up the serial arrangement of animal forms. In 1809, in the second volume of his *Philosophie Zoologique*, as Packard has pointed out, he arranged animals according to their relationships, in the form of a trunk with divergent branches. This was no vague suggestion on his part, but an actual pictorial representation of the relationship between different groups of animals, as conceived by him. Although a crude attempt, it is interesting as being the first of its kind. This is so directly opposed to the idea of scale of being that we make note of the fact that Lamarck forsook that view at least twentv years before the close of his life and substituted for it that of the genealogical tree.

Lamarck's Position in Science.—Lamarck is coming into full recognition for his part in founding the evolution theory, but he is not generally, as yet, given due credit for his work in zoölogy. He was the most philosophical thinker engaged with zoölogy at the close of the eighteenth and the beginning of the nineteenth century. He was greater than Cuvier in his reach of intellect and in his discernment of the true relationships among living organisms. We are to recollect that he forsook the dogma of fixity of species, to which Cuvier held, and founded the first comprehensive theory of organic evolution. To-dav we can recognize the superiority of his mental grasp over that of Cuvier, but, owing to the personal magnetism of the latter and to his position, the ideas of Lamarck, which Cuvier combated, received but little attention when they were promulgated. We shall have occasion in a later chapter to speak more fully of Lamarck's contribution to the progress of biological thought.

Cuvier's Four Branches.—We now return to the type theory of Cuvier. By extended studies in comparative anatomy, he came to the conclusion that animals are constructed upon four distinct plans or types: the vertebrate type; the molluscan type; the articulated type, embracing animals with joints or segments; and the radiated type, the latter with a radial arrangement of parts, like the starfish; etc. These types are distinct, but their representatives, instead of forming a linear series, overlap so that the lowest forms of one of the higher groups are simpler in organization than the higher forms of a lower group. This was very illuminating, and, being founded upon an analysis of structure, was important. It was directly at variance with the idea of scale of being, and overthrew that doctrine.

Cuvier first expressed these views in a pamphlet published in 1795, and later in a better-known paper read before the French Academy in 1812, but for the full development of his type-theory we look to his great volume on the animal kingdom published in 1816. The central idea of his arrangement is contained in the secondary title of his book, "The Animal Kingdom Arranged According to its Organization" (*Le Règne Animal Distribué d'après son Organisation*, 1816). The expression "arranged according to its organization" embraces the feature in which this analysis of animals differs from all previous attempts.

Correlation of Parts.—An important idea, first clearly expressed by Cuvier, was that of correlation of parts. The view that the different parts of an animal are so correlated that a change in one, brought about through changes in use, involves a change in another. For illustration, the cleft hoof is always associated with certain forms of teeth and with the stomach of a ruminant. The sharp claws of flesh-eating animals are associated with sharp, cutting teeth for tearing the flesh of the victims, and with an alimentary tube adapted to the digestion of a fleshy diet. Further account of Cuvier is reserved for the chapter on the Rise of Comparative Anat omy, of which he was the founder.

Von Baer.—The next notable advance affecting natural history came through the work of Von Baer, who, in 1828, founded the science of development of animal forms. He arrived at substantially the same conclusions as Cuvier. Thus the system founded upon comparative anatomy by Cuvier came to have the support of Von Baer's studies in embryology.

The contributions of these men proved to be a turning-point in natural history, and subsequent progress in systematic botany and zoölogy resulted from the application of the methods of Cuvier and Von Baer, rather than from following that of Linnæus. His nomenclature remained a permanent contribution of value, but the knowledge of the nature of living forms has been advanced chiefly by studies in comparative anatomy and embryology, and, also, in the application of experiments.

The most significant advances in reference to the classification of animals was to come as a result of the acceptance of the doctrine of organic evolution, subsequent to 1859. Then the relationships between animals were made to depend upon community of descent, and a distinction was drawn between superficial or apparent relationships and those deep-seated characteristics that depend upon close genetic affinities.

Alterations by Von Siebold and Leuckart.—But, in the mean time, naturalists were not long in discovering that the primary divisions established by Cuvier were not well balanced, and, indeed, that they were not natural divisions of the animal kingdom. The group Radiata was the least sharply defined, since Cuvier had included in it not only those animals which exhibit a radial arrangement of parts, but also unicellular organisms that were asymmetrical, and some of the worms that showed bilateral symmetry. Accordingly, Karl Th. von Siebold, in 1845, separated these animals and redistributed them. For the simplest unicellular animals he adopted the name Protozoa, which they still retain, and the truly radiated forms, as starfish, sea-urchins, hydroid polyps, coral animals, etc., were united in the group Zoöphyta. Von Siebold also changed Cuvier's branch, Articulata, separating those forms as crustacea, insects, spiders, and myriopods, which have jointed appendages, into a natural group called Arthropoda, and uniting the segmented worms with those worms that Cuvier has included in the radiate group, into another branch called Vermes. This separation of the four original branches of Cuvier was a movement in the right direction, and was destined to be carried still farther.

FIG. 35.—KARL TH. VON SIEBOLD, 1804–1885.

Von Siebold (Fig. 35) was an important man in the progress of zoölogy, especially in reference to the comparative anatomy of the invertebrates.

Leuckart (Fig. 36), whose fame as a lecturer and teacher

attracted many young men to the University of Leipsic, is another conspicuous personality in zoölogical progress.

This distinguished zoölogist, following the lead of Von Siebold, made further modifications. He split Von Siebold's group of Zoöphytes into two distinct kinds of radiated animals:

FIG. 36.—RUDOLPH LEUCKART, 1823–1898.

the star-fishes, sea-urchins, sea-cucumbers, etc., having a spiny skin, he designated Echinoderma; the jelly-fishes, polyps, coral animals, etc., not possessing a true body cavity, were also united into a natural group, for which he proposed the name Cœlenterata.

From all these changes there resulted the seven primary

divisions—branches, subkingdoms, or phyla—which, with small modifications, are still in use. These are Protozoa, Cœlenterata, Echinoderma, Vermes, Arthropoda, Mollusca, Vertebrata. These seven phyla are not entirely satisfactory, and there is being carried on a redistribution of forms, as in the case of the brachiopods, the sponges, the tunicates, etc. While all this makes toward progress, the changes are of more narrow compass than those alterations due to Von Siebold and Leuckart.

Summary.—In reviewing the rise of scientific natural history, we observe a steady development from the time of the *Physiologus*, first through a return to Aristotle, and through gradual additions to his observations, notably by Gesner, and then the striking improvements due to Ray and Linnæus. We may speak of the latter two as the founders of systematic botany and zoölogy. But the system left by Linnæus was artificial, and the greatest obvious need was to convert it into a natural system founded upon a knowledge of the structure and the development of living organisms. This was begun by Cuvier and Von Baer, and was continued especially by Von Siebold and Leuckart. To this has been added the study of habits, breeding, and adaptations of organisms, a study which has given to natural history much greater importance than if it stood merely for the systematic classification of animals and plants.

Tabular View of Classifications.—A table showing the primary groups of Linnæus, Cuvier, Von Siebold, and Leuckart will be helpful in picturing to the mind the modifications made in the classification of animals. Such a table is given on the following page.

L. Agassiz, in his famous essay on Classification, reviews in the most scholarly way the various systems of classification. One peculiar feature of Agassiz's philosophy was his adherence to the dogma of the fixity of species. The same year that his essay referred to was published (1859) appeared Darwin's *Origin of Species.* Agassiz, however, was never able to accept the idea of the transformations of species.

Linnæus	Cuvier	Von Siebold	Leuckart
Mammalia	Vertebrata (Embracing five classes: Mammalia, Aves, Reptilia, Batrachia, Pisces.)	Vertebrata (Embracing five classes.)	Vertebrata (Five classes.)
Aves			
Amphibia			
Pisces			
Insecta (Including Crustacea, etc.)	Mollusca	Mollusca	Mollusca
	Articulata.....	{ Arthropoda Vermes }	Arthropoda
			Vermes
Vermes (Including Mollusca and all lower forms.)	Radiata.......	{ Zoöphyta.... Protozoa }	{ Echinoderma Cœlenterata } Protozoa

STEPS IN BIOLOGICAL PROGRESS FROM LINNÆUS TO DARWIN

The period from Linnæus to Darwin is one full of important advances for biology in general. We have considered in this chapter only those features that related to changes in the system of classification, but in the mean time the morphological and the physiological sides of biology were being advanced not only by an accumulation of facts, but by their better analysis. It is an interesting fact that, although during this period the details of the subject were greatly multiplied, progress was relatively straightforward and by a series of steps that can be clearly indicated.

It will be of advantage before the subject is taken up in its parts to give a brief forecast in which the steps of progress can be represented in outline without the confusion arising from the consideration of details. Geddes, in 1898, pointed out the steps in progress, and the account that follows is based upon his lucid analysis.

The Organism.—In the time of Linnæus the attention of naturalists was mainly given to the organism as a whole. Plants and animals were considered from the standpoint of the organism—the external features were largely dealt with, the habitat, the color, and the general appearance—features which characterize the organism as a whole. Linnæus and Jussieu represent this phase of the work, and Buffon the higher type of it. Modern studies in this line are like addition to the *Systema Naturæ.*

Organs.—The first distinct advance came in investigating animals and plants according to their structure. Instead of the complete organism, the organs of which it is composed became the chief subject of analysis. The organism was dissected, the organs were examined broadly, and those of one kind of animal and plant compared with another. This kind of comparative studv centered in Cuvier, who, in the early part of the nineteenth century, founded the science of comparative anatomv of animals, and in Hofmeister, who examined the structure of plants on a basis of broad comparison.

Tissues.—Bichat, the famous contemporary of Cuvier, essayed a deeper level of analysis in directing attention to the tissues that are combined to make up the organs. He dis tinguished twenty-one kinds of tissues by combinations of which the organs are composed. This step laid the foundation for the science of histology, or minute anatomy. Bichat called it general anatomv (*Anatomie Générale*, 1801).

Cells.—Before long it was shown that tissues are not the real units of structure, but that they are composed of microscopic elements called cells. This level of analvsis was not reached until magnifying-lenses were greatly improved—it was a product of a closer scrutiny of nature with improved instruments. The foundation of the work, especially for plants, had been laid by Leeuwenhoek, Malpighi, and Grew.

But when the broad generalization, that all the tissues of animals and plants are composed of cells, was given to the world by Schleiden and Schwann, in 1838–39, the entire organization of living forms took on a new aspect. This was progress in understanding the morphology of animals and plants.

Protoplasm.—With improved microscopes and attention directed to cells, it was not long before the discovery was made that the cells as units of structure contain protoplasm. That this substance is similar in plants and animals and is the seat of all vital activity was determined chiefly by the researches of Max Schultze, published in 1861. Thus step by step, from 1758, the date of the tenth edition of the *Systema Naturæ*, to 1861, there was a progress on the morphological side, passing from the organism as a whole to organs, to tissues, to cells, and finally to protoplasm, the study of which in all its phases is the chief pursuit of biologists.

The physiological side had a parallel development. In the period of Linnæus, the physiology of the organism was investigated by Haller and his school; following him the physiology of organs and tissues was advanced by J. Müller, Bichat, and others. Later, Virchow investigated the physiology of cells, and Claude Bernard the chemical activities of protoplasm.

This set forth in outline will be amplified in the following chapters.

CHAPTER VII

CUVIER AND THE RISE OF COMPARATIVE ANATOMY

AFTER observers like Linnæus and his followers had attained a knowledge of the externals, it was natural that men should turn their attention to the organization or internal structure of living beings, and when the latter kind of investigation became broadly comparative, it blossomed into comparative anatomy. The materials out of which the science of comparative anatomy was constructed had been long accumulating before the advent of Cuvier, but the mass of details had not been organized into a compact science.

As indicated in previous chapters, there had been an increasing number of studies upon the structure of organisms, both plant and animal, and there had resulted some noteworthy monographs. All this work, however, was mainly descriptive, and not comparative. Now and then, the comparing tendency had been shown in isolated writings such as those of Harvey, Malpighi, and others. As early as 1555, Belon had compared the skeleton of the bird with that of the human body "in the same posture and as nearly as possible bone for bone"; but this was merely a faint foreshadowing of what was to be done later in comparing the systems of the more important organs.

We must keep in mind that the study of anatomy embraces not merely the bony framework of animals, but also the muscles, the nervous system, the sense organs, and all the other structures of both animals and plants. In the rise of comparative anatomy there gradually emerged naturalists who compared the structure of the higher animals with that of the simpler ones. These comparisons brought out so many resemblances and so many remarkable facts that anat-

FIG. 37.—SEVERINUS, 1580–1656.

omy, which seems at first a dry subject, became endued with great interest.

Severinus.—The first book expressly devoted to comparative anatomy was that of Severinus (1580–1656), designated

Zootomia Democritæ. The title was derived from the Roman naturalist Democritæus, and the date of its publication, 1645, places the treatise earlier than the works of Malpighi, Leeuwenhoek, and Swammerdam. The book is illustrated by numerous coarse woodcuts, showing the internal organs of fishes, birds, and some mammals. There are also a few illustrations of stages in the development of these animals. The comparisons were superficial and incidental; nevertheless, as the first attempt, after the revival of anatomy, to make the subject comparative, it has some especial interest. Severinus (Fig. 37) should be recognized as beginning the line of comparative anatomists which led up to Cuvier.

Forerunners of Cuvier.—Anatomical studies began to take on broad features with the work of Camper, John Hunter, and Vicq d'Azyr. These three men paved the way for Cuvier, but it must be said of the two former that their comparisons were limited and unsystematic.

Camper, whose portrait is shown in Fig. 38, was born in Leyden, in 1722. He was a versatile man, having a taste for drawing, painting, and sculpture, as well as for scientific studies. He received his scientific training under Boerhaave and other eminent men in Leyden, and became a professor and, later, rector in the University of Groningen. Possessing an ample fortune, and also having married a rich wife, he was in position to follow his own tastes. He travelled extensively and gathered a large collection of skeletons. He showed considerable talent as an anatomist, and he made several discoveries, which, however, he did not develop, but left to others. Perhaps the possession of riches was one of his limitations; at any rate, he lacked fixity of purpose.

Among his discoveries may be mentioned the semicircular canals in the ear of fishes, the fact that the bones of flying birds are permeated by air, the determination of some fossil bones, with the suggestion that they belonged to extinct forms.

The latter point is of interest, as antedating the conclusions of Cuvier regarding the nature of fossil bones. Camper also made observations upon the facial angle as an index of intelligence in the different races of mankind, and in lower

FIG. 38.—CAMPER, 1722–1789.

animals. He studied the anatomy of the elephant, the whale, the orang, etc.

John Hunter (1728–1793), the gifted Scotchman whose museum in London has been so justly celebrated, was a man of extraordinary originality, who read few books but went directly to nature for his facts; and, although he made errors from which he would have been saved by a wider acquaintance with the writings of naturalists, his neglect of reading left his mind unprejudiced by the views of others. He was a wild, unruly spirit, who would not be forced into the conventional mold as regards either education or manners. His older brother, William, a man of more elegance and refinement, who well understood the value of polish in refer-

FIG. 39.—JOHN HUNTER, 1728–1793.

ence to worldly success, tried to improve John by arranging for him to go to the University of Oxford, but John rebelled and would not have the classical education of the university, nor would he take on the refinements of taste and manner of which his brother was a good example. "Why," the doughty John is reported to have said, "they wanted to make me study

Greek! They tried to make an old woman of me!" However much lack of appreciation this attitude indicated, it shows also the Philistine independence of his spirit. This independence of mind is one of his striking characteristics.

This is not the place to dwell upon the unfortunate controversy that arose between these two illustrious brothers regarding scientific discoveries claimed by each. The position of both is secure in the historical development of medicine and surgery. Although the work of John Hunter was largely medical and surgical, he also made extensive studies on the comparative anatomy of animals, and has a place as one of the most conspicuous predecessors of Cuvier. He was very energetic both in making discoveries and in adding to his great museum.

The original collections made by Hunter are still open to inspection in the rooms of the Royal College of Surgeons, London. It was his object to preserve specimens to illustrate the phenomena of life in all organisms, whether in health or disease, and the extent of his museum may be divined from the circumstance that he expended upon it about three hundred and seventy-five thousand dollars. Although he described and compared many types of animals, it was as much in bringing this collection together and leaving it to posterity that he advanced comparative anatomy as in what he wrote. After his death the House of Commons purchased his museum for fifteen thousand pounds, and placed it under the care of the corporation of Surgeons. Hunter's portrait is shown in Fig. 39.

Vicq d'Azyr (Fig. 40), more than any other man, holds the chief rank as a comparative anatomist before the advent of Cuvier into the same field. He was born in 1748, the son of a physician, and went to Paris at the age of seventeen to study medicine, remaining in the metropolis to the time of his death in 1794. He was celebrated as a physician, became permanent secretary of the newly founded Academy of Medicine, consulting physician to the queen, and occupied other positions of trust and responsibility. He married the niece of Daubenton, and, largely through his influence, was advanced to social place and recognition. On the death of Buffon, in 1788, he took the seat of that distinguished naturalist as a member of the French Academy.

FIG. 40.—VICQ D'AZYR, 1748–1794.

He made extensive studies upon the organization particularly of birds and quadrupeds, making comparisons between their structure, and bringing out new points that were superior to anything yet published. His comparisons of the limbs of man and animals, showing a correspondence between the flexor and extensor muscles of the legs and arms, were made with great exactness, and they served to mark the beginning of a new kind of precise comparison. These were not merely fanciful comparisons, but exact ones—part for part; and his general considerations based upon these comparisons were of a brilliant character.

As Huxley has said, "he may be considered as the founder of the modern science of anatomy." His work on the structure of the brain was the most exact which had appeared up to that time, and in his studies on the brain he entered into broad comparisons as he had done in the study of the other parts of the animal organization.

He died at the age of forty-six, without being able to complete a large work on human anatomy, illustrated with colored figures. This work had been announced and entered upon, but only that part relating to the brain had appeared at the time of his death. Besides drawings of the exterior of the brain, he made sections; but he was not able to determine with any particular degree of accuracy the course of fiber tracks in the brain. This was left for other workers. He added many new facts to those of his predecessors, and by introducing exact comparisons in anatomy he opened the field for Cuvier.

Cuvier.—When Cuvier, near the close of the eighteenth century, committed himself definitely to the progress of natural science, he found vast accumulations of separate monographs to build upon, but he undertook to dissect representatives of all the groups of animals, and to found his comparative anatomy on personal observations. The work of Vicq d'Ayzr marked the highest level of attainment, and afforded a good model of what comparisons should be; but Cuvier had even larger ideas in reference to the scope of comparative anatomy than had his great predecessor.

The particular feature of Cuvier's service was that in his investigations he covered the whole field of animal organization from the lowest to the highest, and uniting his results with what had already been accomplished, he established comparative anatomy on broad lines as an independent branch of natural science. Almost at the outset he conceived the idea of making a comprehensive study of the structure of the animal kingdom. It was fortunate that he began his investigations with thorough work upon the invertebrated animals; for from this view-point there was gradually unfolded to his great mind the plan of organization of the entire series of animals. Not only is a knowledge of the structure of the simplest animals an essential in understanding that of the more modified ones, but the more delicate work required in dissecting them gives invaluable training for anatomizing those of more complex construction. The value attached to this part of his training by Cuvier is illustrated by the advice that he gave to a young medical student who brought to his attention a supposed discovery in anatomy. "Are you an entomologist?" inquired Cuvier. "No," said the young man. "Then," replied Cuvier, "go first and anatomize an insect and return to me; and if you still believe that your observations are discoveries I will then believe you."

Birth and Early Education.—Cuvier was born in 1769, at Montbéliard, a village at that time belonging to Württemberg, but now a part of the French Jura. His father was a retired military officer of the Swiss army, and the family, being Protestants, had moved to Montbéliard for freedom from religious persecution. Cuvier was christened Léopold-Christian-Frédéric-Dagobert Cuvier, but early in youth took the name of Georges at the wish of his mother, who had lost an infant son by that name.

He gave an early promise of intellectual leadership, and his mother, although not well educated, took the greatest pains in seeing that he formed habits of industry and continuous work, hearing him recite his lessons in Latin and other branches, although she did not possess a knowledge of Latin. He early showed a leaning toward natural history; having access to the works of Gesner and Buffon, he profited by reading these two writers. So great was his interest that he colored the plates in Buffon's *Natural History* from descriptions in the text.

It was at first contemplated by his family that he should prepare for theology, but failing, through the unfairness of one of his teachers, to get an appointment to the theological seminary, his education was continued in other directions. He was befriended by the sister of the Duke of Württemberg, who sent him as a pensioner to the famous Carolinian academy at Stuttgart. There he showed great application, and with the wonderful memory with which he was endowed, he took high rank as a student. Here he met Kielmeyer, a young instructor only four years older than himself, who shared his taste for natural history and, besides this, introduced him to anatomy. In after-years Cuvier acknowledged the assistance of Kielmeyer in determining his future work and in teaching him to dissect.

Life at the Seashore.—In 1788 the resources of his family, which had always been slender, became further reduced by the inability of the government to pay his father's retiring stipend. As the way did not open for employment in other directions, young Cuvier took the post of instructor of the only son in the family of Count d'Héricy, and went with the family to the sea-coast in Normandy, near Caen. For six years (1788–1794) he lived in this noble family, with much time at his disposal. For Cuvier this period, from the age of nineteen to twenty-five, was one of constant research and reflection.

While Paris was disrupted by the reign of terror, Cuvier, who, although of French descent, regarded himself as a German, was quietly carrying on his researches into the strucure of the life at the seaside. These years of diligent study and freedom from distractions fixed his destiny. Here at the sea-coast, without the assistance of books and the stimulus of intercourse with other naturalists, he was drawn directly to nature, and through his great industry he became an independent observer. Here he laid the foundation of his extensive knowledge of comparative anatomy, and from this quiet spot he sent forth his earliest scientific writings, which served to carry his name to Paris, the great center of scientific research in France.

Goes to Paris.—His removal from these provincial surroundings was mainly owing to the warm support of Tessier, who was spending the time of the reign of terror in retirement in an adjacent village, under an assumed name. He and Cuvier met in a scientific society, where the identity of Tessier was discovered by Cuvier on account of his ease of speech and his great familiarity with the topics discussed. A friendship sprung up between them, and Tessier addressed some of his scientific friends in Paris in the interest of Cuvier. By this powerful introduction, and also through the intervention of Geoffroy Saint-Hilaire, he came to Paris in 1795 and was welcomed into the group of working naturalists at the Jardin des Plantes, little dreaming at the time that he should be the leader of the group of men gathered around this scientific institution. He was modest, and so uncertain of his future that for a year he held to his post of instructor, bringing his young charge with him to Paris.

Notwithstanding the doubt which he entertained regarding his abilities, his career proved successful from the beginning. In Paris he entered upon a brilliant career, which was a succession of triumphs. His unmistakable talent, combined with industry and unusual opportunities, brought him rapidly to the front. The large amount of material already collected, and the stimulating companionship of other scientific workers, afforded an environment in which he grew rapidly. He responded to the stimulus, and developed not only into a great naturalist, but expanded into a finished gentleman of the world. Circumstances shaped themselves so that he was called to occupy prominent offices under the government, and he came ultimately to be the head of the group of scientific men into which he had been welcomed as a young man from the provinces.

His Physiognomy.—It is very interesting to note in his portraits the change in his physiognomy accompanying his transformation from a young man of provincial appearance

FIG. 41.—CUVIER AS A YOUNG MAN, 1769–1832.

into an elegant personage. Fig. 41 shows his portrait in the early days when he was less mindful of his personal appearance. It is the face of an eager, strong, young man, still retaining traces of his provincial life. His long, light-colored hair is unkempt, but does not hide the magnificent proportions of his head. Fig. 42 shows the growing refinement of features which came with his advancement, and the aristocratic look of supremacy which set upon his countenance after his wide recognition passing by a gradation of steps from the position of head of the educational system, to that of baron and peer of France.

Cuvier was a man of commanding power and colosal

FIG. 42.—CUVIER AT THE ZENITH OF HIS POWER.

attainments; he was a favorite of Napoleon Bonaparte, who elevated him to office and made him director of the higher educational institutions of the Empire. But to whatever place of prominence he attained in the government, he never lost his love for natural science. With him this was an absorbing passion, and it may be said that he ranks higher as a zoölogist than as a legislator.

Comprehensiveness of Mind.—Soon after his arrival in Paris he began to lecture upon comparative anatomy and to continue work in a most comprehensive way upon the subjects which he had cultivated at Caen. He saw everything on a large scale. This led to his making extensive studies of whatever problems engaged his mind, and his studies were combined in such a manner as to give a broad view of the subject.

Indeed, comprehensiveness of mind seems to have been the characteristic which most impressed those who were acquainted with him. Flourens says of him: "*Ce qui caractérise partout M. Cuvier, c'est l'esprit vaste.*" His broad and comprehensive mind enabled him to map out on great lines the subject of comparative anatomy. His breadth was at times his undoing, for it must be confessed that when the details of the subject are considered, he was often inaccurate. This was possibly owing to the conditions under which he worked; having his mind diverted into many other channels, never neglecting his state duties, it is reasonable to suppose that he lacked the necessary time to prove his observations in anatomy, and we may in this way account for some of his inaccuracies.

Besides being at fault in some of his comparative anatomy, he adhered to a number of ideas that served to retard the progress of science. He was opposed to the ideas of his contemporary Lamarck, on the evolution of animals. He is remembered as the author of the dogma of catastrophism in geology. He adhered to the old notion of the pre-formation of the embryo, and also to the theory of the spontaneous origin of life.

Founds Comparative Anatomy.—Regardless of this qualification, he was a great and distinguished student, and founded comparative anatomy. From 1801 to 1805 appeared his *Leçons d'Anatomie Comparée*, a systematic treatise on the comparative anatomy of animals, embracing both the invertebrates and the vertebrates. In 1812 was published his great work on the fossil bones about Paris, an achievement which founded the science of vertebrate palæontology. His extensive examination of the structure of fishes also added to his already great reputation. His book on the animal kingdom (*Le Règne Animal distribué d'après son Organisation*, 1816), in which he expounded his type-theory, has been considered in a previous chapter.

He was also deeply interested in the historical development of science, and his volumes on the rise of the natural sciences give us almost the best historical estimate of the progress of science that we have at the present day.

His Domestic Life.—Mrs. Lee, in a chatty account of Cuvier, shows one of his methods of work. He had the faculty of making others assist him in various ways. Not only members of his family, but also guests in his household were pressed into service. They were invited to examine different editions of works and to indicate the differences in the plates and in the text. This practice resulted in saving much time for Cuvier, since in the preparation of his historical lectures he undertook to examine all the original sources of the history with which he was engaged. In his lectures he summarized facts relating to different editions of books, etc.

Mrs. Lee also gives a picture of his family life, which was, to all accounts, very beautiful. He was devoted to his wife and children, and in the midst of exacting cares he found time to bind his family in love and devotion. Cuvier was called upon to suffer poignant grief in the loss of his children, and his direct family was not continued. He was especially broken by the death of his daughter who had grown to young womanhood and was about to be married.

From the standpoint of a sincere admirer, Mrs. Lee writes of his generosity and nobility of temperament, declaring that his career demonstrated that his mind was great and free from both envy and smallness.

Some Shortcomings.—Nevertheless, there are certain things in the life of Cuvier that we wish might not have been. His break with his old friends Lamarck and Saint-Hilaire seems to show a domination of qualities that were not generous and kindly; those observations of Lamarck showing a much profounder insight than any of which he himself was the author were laughed to scorn. His famous controversy with Saint-Hilaire marks a historical moment that will be dealt with in the chapter on Evolution.

George Bancroft, the American historian, met him during a visit to Paris in 1827. He speaks of his magnificent eyes and his fine appearance, but on the whole Cuvier seems to have impressed Bancroft as a disagreeable man.

Some of his shortcomings that served to retard the progress of science have been mentioned. Still, with all his faults, he dominated zoölogical science at the beginning of the nineteenth century, and so powerful was his influence and so undisputed was his authority among the French people that the rising young men in natural science sided with Cuvier even when he was wrong. It is a noteworthy fact that France, under the influence of the traditions of Cuvier, was the last country slowly and reluctantly to harbor as true the ideas regarding the evolution of animal life.

Cuvier's Successors

While Cuvier's theoretical conclusions exercised a retarding influence upon the progress of biology, his practical studies more than compensated for this. It has been pointed out how his type-theory led to the reform of the Linnæan system, but, besides this, the stimulus which his investigations gave to studies in comparative anatomy was even of more beneficent influence. As time passed the importance of comparative anatomy as one division of biological science impressed itself more and more upon naturalists. A large number of investigators in France, England, and Germany entered the field and took up the work where Cuvier had

FIG. 43.—H. MILNE-EDWARDS, 1800–1885.

left it. The more notable of these successors of Cuvier should come under consideration.

His intellectual heirs in France were Milne-Edwards and Lacaze-Duthiers.

Milne-Edwards.—H. Milne-Edwards (1800–1885) was a man of great industry and fine attainments; prominent alike in comparative anatomy, comparative physiology, and general zoölogy, professor for many years at the Sorbonne in Paris.

In 1827 he introduced into biology the fruitful idea of the division of physiological labor. He completed and published excellent researches upon the structure and development of many animals, notably crustacea, corals, etc. His work on comparative anatomy took the form of explanations of the activities of animals, or comparative physiology. His comprehensive treatise *Leçons sur la Physiologie et l'Anatomie Comparée*, in fourteen volumes, 1857–1881, is a mine of information regarding comparative anatomy as well as the physiology of organisms.

Lacaze-Duthiers.—Henri de Lacaze-Duthiers (1821–1901), the man of comprehensive mind, stimulating as an instructor of young men, inspiring other workers, and producing a large amount of original research on his own account, director of the Seaside Stations at Roscoff and Banyuls, the founder of a noteworthy periodical of experimental zoölogy—this great man, whose portrait is shown in Fig. 44, was one of the leading comparative anatomists in France.

R. Owen.—In England Richard Owen (1804–1892) carried on the influence of Cuvier. At the age of twenty-seven he went to Paris and renewed acquaintance with the great Cuvier, whom he had met the previous year in England. He spent some time at the Jardin des Plantes examining the extensive collections in the museum. Although the idea was repudiated by Owen and some of his friends, it is not unlikely that the collections of fossil animals and the researches upon them which engaged Cuvier at that time had great influence upon the subsequent studies of Owen. Although he never studied under Cuvier, in a sense he may be regarded as his disciple. Owen introduced into anatomy the important conceptions of analogy and homology, the former being a likeness based upon the use to which organs are put, as the wing of a butterfly and the wing of a bat; while homology is a true relationship founded on likeness in structure and development, as the wing of a bat and the foreleg of a dog. Analogy is a superficial, and often a deceiving relationship; homology is a true genetic relationship. It is obvious that this distinction

FIG. 44.—LACAZE-DUTHIERS, 1821–1901.

is of great importance in comparing the different parts of animals. He made a large number of independent discoveries, and published a monumental work on the comparative anatomy of vertebrates (1866–68). In much of his thought he was singular, and manv of his general conclusions have not stood the test of time. He undertook to establish the idea of an archtype in vertebrate anatomy. He clung to the vertebral theory of the skull long after Huxley had shown such a theory to be untenable. The idea that the skull is made up

FIG. 45.—LORENZO OKEN, 1779-1851.

of modified vertebræ was propounded by Goethe and Oken. In the hands of Oken it became one of the anatomical conclusions of the school of *Naturphilosophie.* This school of transcendental philosophy was founded by Schelling, and Oken (Fig. 45) was one of its typical representatives. The vertebral theory of the skull was, therefore, not original with Owen, but he adopted it, greatly elaborated it, and clung to it blindly long after the foundations upon which it rested were removed.

Richard Owen (Fig. 46) was succeeded by Huxley (1825–1895), whose exactness of observation and rare judgment

FIG. 46.—RICHARD OWEN, 1804–1892.

as to the main facts of comparative anatomy mark him as one of the leaders in this field of research. The influence of Huxley as a popular exponent of science is dealt with in a later chapter.

Meckel.—Just as Cuvier stands at the beginning of the school of comparative anatomy in France, so does J. Fr. Meckel in Germany. Meckel (1781–1833) was a man of rare talent, descended from a family of distinguished anatomists. From 1804 to 1806 he studied in Paris under Cuvier, and when he came to leave the French capital to become professor of anatomy at Halle, he carried into Germany the

FIG. 47.—J. FR. MECKEL, 1781–1833.

teachings and methods of his master. He was a strong force in the university, attracting students to his department by his excellent lectures and his ability to arouse enthusiasm. Some of these students were stimulated to undertake researches in anatomy, and there came from his laboratory a number of investigations that were published in a periodical which he founded. Meckel himself produced many scientific papers and works on comparative anatomy, which assisted materially in the advancement of that science. His portrait, which is rare, is shown in Fig. 47.

Rathke.—Martin Henrv Rathke (1793–1860) greatly advanced the science of comparative anatomy by insisting upon the importance of elucidating anatomy with researches in development. This is such an important consideration that his influence upon the progress of comparative anatomy can not be overlooked. After being a professor in Dorpat, he came, in 1835, to occupy the position of professor of anatomy and zoölogy at Königsberg, which had been vacated by Von Baer on there moval of the latter to St. Petersburg. His writings are composed with great intelligence, and his facts are carefully coördinated. Rathke belonged to the good old school of German writers whose researches were profound and extensive, and whose expression was clear, being based upon matured thought. His papers on the aortic arches and the Wolffian bodv are those most commonly referred to at the present time.

Müller.—Johannes Müller (1801–1858), that phenomenal man, besides securing recognition as the greatest physiologist of the nineteenth century, also gave attention to comparative anatomy, and earned the title of the greatest morphologist of his time. His researches were so accurate, so complete, so discerning, that his influence upon the development of comparative anatomy was profound. Although he is accorded, in history, the double distinction of being a great anatomist and a great physiologist, his teaching tended to physiology; and most of his distinguished students were physiologists of the broadest tvpe, uniting comparative anatomy with their researches upon functional activities. (For Müller's portrait see p. 187.)

Gegenbaur.—In Karl Gegenbaur (1826–1903) scientific anatomy reached its highest expression. His work was characterized by broad and masterly analvsis of the facts of strueture, to which were added the ideas derived from the study of the development of organs. He was endowed with an intensely keen insight, an insight which enabled him to separate from the vast mass of facts the important and essential features, so that they yielded results of great interest and of lasting importance. This gifted anatomist attracted many young men

FIG. 48.—KARL GEGENBAUR, 1826–1903.

from the United States and from other countries to pursue under his direction the study of comparative anatomy. He died in Heidelberg in 1903, where he had been for many years professor of anatomy in the university.

In the group of living German anatomists the names of Fürbringer, Waldeyer, and Wiedersheim can not go unmentioned.

E. D. Cope.—In America the greatest comparative anatomist was E. D. Cope (1840–1897), a man of the highest order of attainment, who dealt with the comparative anatomy not only of living forms, but of fossil life, and made contributions of a permanent character to this great science; a man whose title to distinction in the field of comparative anatomy will become clearer to later students with the passage of time. For Cope's portrait see p. 336.

Of the successors of Cuvier, we would designate Meckel, Owen, Gegenbaur, and Cope as the greatest.

Comparative anatomy is a very rich subject, and when elucidated by embryology, is one of the firm foundations of biology. If we regard anatomy as a science of statics, we recognize that it should be united with physiology, which represents the dynamical side of life. Comparative anatomy and comparative physiology should go hand in hand in the attempt to interpret living forms. Advances in these two subjects embrace nearly all our knowledge of living organisms. It is a cause for congratulation that comparative anatomy has now become experimental, and that gratifying progress is being made along the line of research designated as experimental morphology. Already valuable results have been attained in this field, and the outlook of experimental morphology is most promising.

CHAPTER VIII

BICHAT AND THE BIRTH OF HISTOLOGY

We must recognize Bichat as one of the foremost men in biological history, although his name is not well known to the general public, nor constantly referred to by biologists as that of one of the chief luminaries of their science. In him was combined extraordinary talent with powers of intense and prolonged application; a combination which has always produced notable results in the world. He died at the age of thirty-one, but, within a productive period of not more than seven years, he made observations and published work that created an epoch and made a lasting impression on biological history.

His researches supplemented those of Cuvier, and carried the analysis of animal organization to a deeper level. Cuvier laid the foundations of comparative anatomy by dissecting and arranging in a comprehensive system the organs of animals, but Bichat went a step further and made a profound study of the tissues that unite to make up the organs. As we have already noted in a previous chapter, this was a step in reaching the conception of the real organization of living beings.

Buckle's Estimate of Bichat.—It is interesting to note the impression made by Bichat upon one of the greatest students of the history of civilization. Buckle says of him· "Great, however, as is the name of Cuvier, a greater still remains behind. I allude, of course, to Bichat, whose reputation is steadily advancing as our knowledge advances; who, if we compare the shortness of his life with the reach and depth of his views, must be pronounced the most profound thinker and consummate observer by whom the organization of the animal frame has yet been studied.

"We may except Aristotle, but between Aristotle and Bichat I find no middle man."

Whether or not we agree fully with this panegyric of Buckle, we must, I think, place Bichat among the most illus trious men of biological history, as Vesalius, J. Müller, Von Baer, and Balfour.

Marie François Xavier Bichat was born in 1771 at Thoirette, department of the Ain. His father, who was a physician, directed the early education of his son and had the satisfaction of seeing him take kindly to intellectual pursuits. The young student was distinguished in Latin and mathematics, and showed early a fondness for natural history. Having elected to follow the calling of his father, he went to Lyons to study medicine, and came under the instruction of Petit in surgery.

Bichat in Paris.—It was, on the whole, a fortunate circumstance for Bichat that the turbulent events of the French Revolution drove him from Lyons to Paris, where he could have the best training, the greatest stimulus for his growth, and at the same time the widest field for the exercise of his talents. We find him in Paris in 1793, studying under the great surgeon Desault.

He attracted attention to himself in the class of this distinguished teacher and operator by an extemporaneous report on one of the lectures. It was the custom in Desault's classes to have the lectures of the professor reported upon before an assistant bv some student especially appointed for the purpose. On one occasion the student who had been appointed to prepare and deliver the review was absent, and Bichat, who was gifted with a powerful memory, volunteered without previous notice to take his place. The lecture was a long and difficult one on the fractures of the clavicle, but Bichat's abstract was so clear, forceful, and complete that its delivery in well-chosen language produced a great sensation both upon the instructor and the students. This notable performance served to bring him directly to the attention of Desault, who invited him to become his assistant and to live in his family. The association of Bichat with the great surgeon was most happy. Desault treated him as a son, and when he suddenly died in 1795, the care of preparing his works for the printer was left to Bichat.

The fidelity with which Bichat executed this trust was characteristic of his noble nature. He laid aside his own personal interests, and his researches in which he was already immersed, and by almost superhuman labor completed the fourth volume of Desault's *Journal of Surgery* and at the same time collected and published his scattered papers. To these he added observations of his own, making alterations to bring the work up to the highest plane. Thus he paid the debt of gratitude which he felt he owed to Desault for his friendship and assistance.

In 1797 he was appointed professor of anatomy, at the age of twenty-six, and from then to the end of his life, in 1801, he continued in his career of remarkable industry.

The portrait of this very attractive man is shown in Fig. 49. His face shows strong intellectuality. He is described as of "middling stature, with an agreeable face lighted by piercing and expressive eyes." He was much beloved by his students and associates, being "in all relations of life most amiable, a stranger to envy or other hateful passions, modest in demeanor and lively in his manners, which were open and free."

His Phenomenal Industry.—His industry was phenomenal; besides doing the work of a professor, he attended to a considerable practice, and during a single winter he is said to have examined with care six hundred bodies in the pursuance of his researches upon pathological anatomy.

FIG. 49.—BICHAT, 1771–1801.

In the year 1800, when he was thirty years old, began to appear the results of his matured researches. We speak of these as being matured, not on account of his age or the great number of years he had labored upon them, but from the intensity and completeness with which he had pursued his investigations, thus giving to his work a lasting quality.

First came his treatise on the membranes (*Traité des Membranes*); followed quickly by his Physiological Researches into the Phenomena of Life and Death (*Recherches Physiologiques sur la Vie et la Mort*); then appeared his General Anatomy (*Anatomie Générale*) in 1801, and his treatise upon Descriptive Anatomy, upon which he was working at the time of his death.

His death occurred in 1801, and was due partly to an accident. He slipped upon the stairs of the dissecting-room, and his fall was followed by gastric derangement, from which he died.

Results of His Work.—The new science of the anatomy of the tissues which he founded is now known as histology, and the general anatomy, as he called it, has now become the study of minute anatomy of the tissues. Bichat studied the membranes or tissues very profoundly, but he did not employ the microscope and make sketches of their cellular construction. The result of his work was to set the world studying the minute structure of the tissues, a consequence of which led to the modern study of histology. Since this science was constructed directly upon his foundation, it is proper to recognize him as the founder of histology.

Carpenter says of him: "Altogether Bichat left an impress upon the science of life, the depth of which can scarcely be overrated; and this not so much by the facts which he collected and generalized, as by the method of inquiry which he developed, and by the systematic form which he gave to the study of general anatomy in its relations both to physiology and pathology."

Bichat's More Notable Successors.—His influence extended far, and after the establishment of the cell-theory took on a new phase. Microscopic study of the tissues has now become a separate division of the science of anatomy, and engages the attention of a very large number of workers. While the men who built upon Bichat's foundation are numerous, we shall select for especial mention only a few of the more notable, as Schwann, Koelliker, Schultze, Virchow, Leydig, and Ramon y Cajal, whose researches stand in the direct line of development of the ideas promulgated by Bichat.

Schwann.—Schwann's cell-theory was the result of close attention to the microscopic structure of the tissues of animals. It was an extension of the knowledge of the tissues which Bichat distinguished and so thoroughly investigated from other points of view. The cell-theory, which took rise in 1839, was itself epoch-making, and the science of general anatomy was influenced by it as deeply as was the science of embryology. The leading founder of this theory was Theodor Schwann, whose portrait is shown on page 245, where there is also a more extended account of his labors in connection with the cell-theory. Had not the life of Bichat been cut off in his early manhood, he might well have lived to see this great discovery added to his own.

Koelliker.—Albrecht von Koelliker (1817–1905) was one of the greatest histologists of the nineteenth century. He is a striking figure in the development of biology in a general way, distinguished as an embryologist, as a histologist, and in other connections. During his long life, from 1817 to 1905, he made an astounding number of additions to our knowledge of microscopic anatomy. In the early years of his scientific activity, "he helped in establishing the cell-theory, he traced the origin of tissues from the segmenting ovum through the developing embryo, he demonstrated the continuity between nerve-fibers and nerve-cells of vertebrates (1845), . . . and much more." He is mentioned further, in connection with the rise of embryology, in Chapter X.

The strong features of this veteran of research are shown in the portrait, Fig. 50, which represents him at the age of seventy.

In 1847 he was called to the University of Würzburg, where he remained to the time of his death. From 1850 to 1900, scarcely a year passed without some important contribution from Von Koelliker extending the knowledge of histology. His famous text-book on the structure of the tissues (*Handbuch der Gewebelehre*) passed through six editions from 1852 to 1893, the final edition of it being worked over and brought up to date by this extraordinary man after he had passed the age of seventy-five. By workers in biology this will be recognized as a colossal task. In the second volume of the last edition of this work, which appeared in 1893, he went completely over the ground of the vast accumulation of information regarding the nervous system which an army of gifted and energetic workers had produced. This was all thoroughly digested, and his histological work brought down to date.

Schultze.—The fine observations of Max Schultze (1825–1874) may also be grouped with those of the histologists. We shall have occasion to speak of him more particularly in the chapter on Protoplasm. He did memorable service for general biology in establishing the protoplasm doctrine, but many of his scientific memoirs are in the line of normal histology; as, those on the structure of the olfactory membrane, on the retina of the eye, the muscle elements, the nerves, etc., etc.

Normal Histology and Pathology.—But histology has two phases: the investigation of the tissues in health, which is called normal histology; and the study of the tissues in disease and under abnormal conditions of development, which is designated pathological histology. The latter division, on account of its importance to the medical man, has

Fig. 50.—Von Koelliker, 1817–1905.

been extensively cultivated, and the development of pathological study has greatly extended the knowledge of the tissues and has had its influence upon the progress of normal histology. Goodsir, in England, and Henle, in Germany, entered the field of pathological histology, both doing work

FIG. 51.—RUDOLPH VIRCHOW, 1821–1903.

of historical importance. They were soon followed by Virchow, whose eminence as a man and a scientist has made his name familiar to people in general.

Virchow.—Rudolph Virchow (1821–1903), for many years a professor in the University of Berlin, was a notable man in biological science and also as a member of the German parliament. He assisted in molding the cell-theory into better form, and in 1858 published a work on *Cellular Pathology*, which applied the cell-theory to diseased tissues. It is to be remembered that Bichat was a medical man, intensely interested in pathological, or diseased, tissues, and we

FIG. 52.—FRANZ LEYDIG, 1821–1908 (April).
Courtesy of Dr. Wm. M. Wheeler.

see in Virchow the one who especially extended Bichat's work on the side of abnormal histology. Virchow's name is associated also with the beginning of the idea of germinal continuity, which is the basis of biological ideas regarding heredity (see, further, Chapter XV).

Leydig.—Franz Leydig (Fig. 52) was early in the field as a histologist with his handbook (*Lehrbuch der Histologie* *des Menschen und der Thiere*) published in 1857. He applied histology especially to the tissues of insects in 1864 and subsequent years, an account of which has already been given in Chapter V.

Cajal as Histologist.—Ramon y Cajal, professor in the University of Madrid, is a histologist whose work in a special

FIG. 53.—S. RAMON Y CAJAL, 1850

field of research is of world-wide renown. His investigations into the microscopic texture of the nervous system and sense-organs have in large part cleared up the questions of the complicated relations between the nervous elements. In company with other European investigators he visited the United States in 1899 on the invitation of Clark University, where his lectures were a feature of the celebration of the tenth anniversary of that university. Besides receiving many honors in previous years, in 1906 he was awarded, in conjunction with the Italian histologist Golgi, one of the Nobel prizes in recognition of his notable investigations. Golgi invented the staining methods that Ramon y Cajal has applied so extensively and so successfully to the histology of the nervous system.

These men in particular may be remembered as the investigators who expanded the work of Bichat on the tissues: Schwann, for disclosing the microscopic elements of animal tissues and founding the cell-theory; Koelliker, as the typical histologist after the analysis of tissues into their elementary parts; Virchow, as extending the cell-idea to abnormal histology; Leydig, for applying histology to the lower animals; and Ramon y Cajal, for investigations into the histology of the nervous system.

Text-Books of Histology.—Besides the works mentioned, the text-books of Frey, Stricker, Ranvier, Klein, Schäfer, and others represent a period in the general introduction of histology to students between 1859 and 1885. But these excellent text-books have been largely superseded by the more recent ones of Stöhr, Boem-Davidoff, Piersol, Szymonowicz, and others. The number of living investigators in histology is enormous; and their work in the subject of cell-structure and in the department of embryology now overlaps.

In pathological histology may be observed an illustration of the application of biological studies to medicine. While no attempt is made to give an account of these practical applications, they are of too great importance to go unmentioned. Histological methods are in constant use in clinical diagnosis, as in blood counts, the study of inflammations, of the action of phagocytes, and of all manner of abnormal growths.

In attempting to trace the beginning of a definite foundation for the work on the structure of tissues, we go back to Bichat rather than to Leeuwenhoek, as Richardson has proposed. Bichat was the first to give a scientific basis for histology founded on extensive observations, since all earlier observers gave only separated accounts of the structure of particular tissues.

CHAPTER IX

THE RISE OF PHYSIOLOGY

HARVEY HALLER JOHANNES MULLER

PHYSIOLOGY had a parallel development with anatomy, but for convenience it will be considered separately. Anatomy shows us that animals and plants are wonderfully constructed, but after we understand their architecture and even their minute structure, the questions remain, What are all the organs and tissues for? and what takes place within the parts that are actually alive? Physiology attempts to answer questions of this nature. It stands, therefore, in contrast with anatomy, and is supplementary to it. The activities of living organisms are varied, and depend on life for their manifestations. These manifestations may be called vital activities. Physiology embraces a study of them all.

Physiology of the Ancients.—This subject began to attract the attention of ancient medical men who wished to fathom the activities of the body in order to heal its diseases, but it is such a difficult thing to begin to comprehend the activities of life that even the simpler relationships were imperfectly understood, and they resorted to mythical explanations. They spoke of spirits and humors in the body as causes of various changes; the arteries were supposed to carry air, the veins only blood; and nothing was known of the circulation. There arose among these early medical men the idea that the body was dominated by a subtle spirit. This went under the name *pneuma*, and the pneuma-theory held sway until the period of the Revival of Learning.

Among the ancient physiologists the great Roman physician Galen is the most noteworthy figure. As he was the greatest anatomist, so he was also the greatest physiologist of ancient times. All physiological knowledge of the time centered in his writings, and these were the standards of physiology for many centuries, as they were also for anatomy. In the early days anatomy, physiology, and medicine were all united into a poorly digested mass of facts and fancies. This state of affairs lasted until the sixteenth century, and then the awakening came, through the efforts of gifted men, endued with the spirit of independent investigation. The advances made depended upon the work or leadership of these men, and there are certain periods of especial importance for the advance of physiology that must be pointed out.

Period of Harvey.—The first of these epochs to be especially noted here is the period of Harvey (1578–1657). In his time the old idea of spirits and humors was giving way, but there was still much vagueness regarding the activities of the body. He helped to illuminate the subject by showing a connection between arteries and veins, and by demonstrating the circulation of the blood. As we have seen in an earlier chapter, Harvey did not observe the blood passing through the capillaries from arteries to veins, but his reasoning was unassailable that such a connection must exist, and that the blood made a complete circulation. He gave his conclusions in his medical lectures as early as 1619, but did not publish his views until 1628. It was reserved for Malpighi, in 1661, actually to see the circulation through the capillaries under the microscope, and for Leeuwenhoek, in 1669 and later years, to extend these observations.

It was during Harvey's life that the microscope was brought into use and was of such great assistance in advancing knowledge. Harvey himself, however, made little use of this instrument. It was during his life also that the knowledge of development was greatly promoted, first through his own efforts, and later through those of Malpighi.

Harvey is to be recognized, then, as the father of modern physiology. Indeed, before his time physiology as such can hardly be spoken of as having come into existence. He introduced experimental work into physiology, and thus laid the foundation of modern investigation. It was the method of Harvey that made definite progress in this line possible, and accordingly we honor him as one of the greatest as well as the earliest of physiologists.

Period of Haller.—From Harvey's time we pass to the period of Haller (1708–1777), at the beginning of which physiology was still wrapped up with medicine and anatomy The great work of Haller was to create an independent science of physiology. He made it a subject to be studied for its own sake, and not merely as an adjunct to medicine. Haller was a man of vast and varied learning, and to him was applied by unsympathetic critics the title of "that abyss of learning." His portrait, as shown in Fig. 54, gives the impression of a somewhat pompous and overbearing personality. He was egotistical, self-complacent, and possessed of great self-esteem. The assurance in the inerrancy of his own conclusions was a marked characteristic of Haller's mind. While he was a good observer, his own work showing conscientious care in observation, he was not a good interpreter, and we are to recollect that he vigorously opposed the idea of development set forth by Wolff, and we must also recognize that his researches formed the chief starting-point of an erroneous conception of vitality.

As Verworn points out, Haller's own experiments upon the phenomena of irritability were exact, but they were misinterpreted by his followers, and through the molding influence of others the attempted explanation of their meaning grew into the conception of a special vital force belonging to living organisms only. In its most complete form, this idea provided for a distinct dualism between living and lifeless matter, making all vital actions dependent upon the

FIG. 54.—ALBRECHT HALLER, 1708–1777.

operation of a mystical supernatural agency. This assumption removed vital phenomena from the domain of clear scientific analysis, and for a long time exercised a retarding influence upon the progress of physiology.

His chief service of permanent value was that he brought into one work all the facts and the chief theories of physiology carefully arranged and digested. This, as has been said, made physiology an independent branch of science, to be pursued for itself and not merely as an adjunct to the study of medicine. The work referred to is his Elements of Physiology (*Elementa Physiologiæ Corporis Humani*, 1758), one of the noteworthy books marking a distinct epoch in the progress of science.

To the period of Haller also belongs the discovery of oxygen, in 1774, by Priestley, a discovery which was destined to have profound influence upon the subsequent development of physiology, so that even now physiology consists largely in tracing the way in which oxygen enters the body, the manner in which it is distributed to the tissues, and the various phases of vital activity that it brings about within the living tissues.

Charles Bell.—The period of Haller may be considered as extending beyond his lifetime and as terminating when the influence of Müller began to be felt. Another discovery coming in the closing years of Haller's period marks a capital advance in physiology. I refer to the discovery of Charles Bell (1774–1842) showing that the nerve fibers of the anterior roots of the spinal cord belong to the motor type, while those of the posterior roots belong to the sensory type.

This great truth was arrived at theoretically, rather than as the result of experimental demonstration. It was first expounded by Bell in 1811 in a small essay entitled *Idea of a New Anatomy of the Brain*, which was printed for private distribution. It was expanded in his papers, beginning in 1821, and published in the Philosophical Transactions of the Royal Society of London, and finally embodied in his work on the nervous system, published in 1830. At this latter date Johannes Müller had reached the age of twenty-nine, and had already entered upon his career as the leading physiologist of Germany. What Bell had divined he demonstrated by experiments.

Charles Bell (Fig. 55) was a surgeon of eminence; in private life he was distinguished by "unpretending amenity, and simplicity of manners and deportment."

FIG. 55.—CHARLES BELL, 1774–1842.

Period of Johannes Müller.—The period that marks the beginning of modern physiology came next, and was due to the genius and force of Johannes Müller (1801–1858). Verworn says of him: "He is one of those monumental figures that the history of every science brings forth but once.

They change the whole aspect of the field in which they work, and all later growth is influenced by their labors." Johannes Müller was a man of very unusual talent and attainments, the possessor of a master mind. Some have said, and not without reason, that there was something supernatural about Müller, for his whole appearance bore the stamp of the uncommon. His portrait, with its massive head above the broad shoulders, is shown in Fig. 56. In his lectures his manner and his gestures reminded one of a Catholic priest. Early in his life, before the disposition to devote himself to science became so overwhelming, he thought of entering the priesthood, and there clung to him all his life some marks of the holy profession. In his highly intellectual face we find "a trace of severity in his mouth and compressed lips, with the expression of most earnest thought on his brow and eyes, and with the remembrance of a finished work in every wrinkle of his countenance."

This extraordinary man exercised a profound influence upon those who came into contact with him. He excited almost unbounded enthusiasm and great veneration among his students. They were allowed to work close by his side, and so magnetic was his personality that he stimulated them powerfully and succeeded in transmitting to them some of his own mental qualities. As professor of physiology in Berlin, Müller trained many gifted young men, among whom were Ludwig (1816–1895), Du Bois-Reymond (1818–1896), and Helmholtz (1821–1894), who became distinguished scholars and professors in German universities. Helmholtz, speaking of Müller's influence on students, paid this tribute to the grandeur of his teacher: "Whoever comes into contact with men of the first rank has an altered scale of values in life. Such intellectual contact is the most interesting event that life can offer."

The particular service of Johannes Müller to science was

to make physiology broadly comparative. So comprehensive was his grasp upon the subject that he gained for himself the title of the greatest physiologist of modern times. He brought together in his great work on the physiology of man not only all that had been previously made known, carefully sifted and digested, but a great mass of new information, which was the result of his own investigations and of those of his students. So rigorous were his scientific standards that he did not admit into this treatise anything which had been untested either by himself or by some of his assistants or students. Verworn says of this monumental work, which appeared in 1833, under the title *Handbuch der Physiologie des Menschen:* "This work stands to-day unsurpassed in the genuinely philosophical manner in which the material, swollen to vast proportions by innumerable special researches, was for the first time sifted and elaborated into a unitary picture of the mechanism within the living organism. In this respect the *Handbuch* is to-day not only unsurpassed, but unequalled."

Müller was the most accurate of observers; indeed, he is the most conspicuous example in the nineteenth century of a man who accomplished a prodigious amount of work all of which was of the highest quality. In physiology he stood on broader lines than had ever been used before. He employed every means at his command—experimenting, the observation of simple animals, the microscope, the discoveries in physics, in chemistry, and in psychology.

He also introduced into physiology the principles of psychology, and it is from the period of Johannes Müller that we are to associate recognition of the close connection between the operations of the mind and the physiology of the brain that has come to occupy such a conspicuous position at the present time.

Müller died in 1858, having reached the age of fifty-seven,

FIG. 56.—JOHANNES MÜLLER, 1801–1858.

but his influence was prolonged through the teachings of his students.

Physiology after Müller

Ludwig.—Among the men who handed on the torch of Müller there has already been mentioned Ludwig (Fig.

Fig. 57.—Ludwig, 1816–1895.

57). For many years he lectured in the University of Leipsic, attracting to that university high-minded, eager, and gifted young men, who received from this great luminary of physiology by expression what he himself had derived from contact with Müller. There are to-day distributed through the universities a number of young

FIG. 58.—DU BOIS-REYMOND, 1818–1896.

physiologists who stand only one generation removed from Johannes Müller, and who still labor in the spirit that was introduced into this department of study by that great master.

Du Bois-Reymond.—Du Bois-Reymond (Fig. 58), another of his distinguished pupils, came to occupy the chair which Müller himself had filled in the University of Berlin, and during the period of his vigor was in physiology one of the lights of the world. It is no uncommon thing to find recently published physiologies dedicated either to the memory of Johannes Müller, as in the case of that remarkable *General Physiology* by Verworn; or to Ludwig, or to Du Bois-Reymond, who were in part his intellectual product. From this disposition among physiologists to do homage to Müller, we are able to estimate somewhat more closely the tremendous reach of his influence.

Bernard.—When Müller was twelve years old there was born in Saint-Julien, department of the Rhône, Claude Bernard, who attained an eminence as a physiologist, of which the French nation are justly proud. Although he was little thought of as a student, nevertheless after he came under the influence of Magendie, at the age of twenty-six, he developed rapidly and showed his true metal. He exhibited great manual dexterity in performing experiments, and also a luminous quality of mind in interpreting his observations. One of his greatest achievements in physiology was the discovery of the formation within the liver of glycogen, a substance chemically related to sugar. Later he discovered the system of vaso-motor nerves that control and regulate the caliber of the blood-vessels. Both of these discoveries assisted materially in understanding the wonderful changes that are going on within the human body. But besides his technical researches, any special consideration of which lies quite beyond the purpose of this book, he published in 1878–1879 a work upon the phenomena of life in animals and vegetables, a work that had general influence in extending the knowledge of vital activities. I refer to his now classic *Leçons sur les Phénomènes de la vie communs aux animaux et aux végétaux.*

The thoughtful face of Bernard is shown in his portrait,

Fig. 59. He was one of those retiring, silent men whose natures are difficult to fathom, and who are so frequentlv misunderstood. A domestic infelicity, that led to the separation of himself from his family, added to his isolation and loneliness. When touched by the social spirit he charmed

FIG. 59.—CLAUDE BERNARD, 1813–1878.

people by his personality. He was admired by the Emperor Napoleon Third, through whose influence Bernard acquired two fine laboratories. In 1868 he was elected to the French Academy, and became thereby one of the "Forty Immortals."

Foster describes him thus: "Tall in stature, with a fine

presence, with a noble head, the eyes full at once of thought and kindness, he drew the look of observers upon him wherever he appeared. As he walked in the streets passers-by might be heard to say ' I wonder who that is; he must be some distinguished man.' "

Two Directions of Growth.—Physiology, established on the broad foundations of Müller, developed along two independent pathways, the physical and the chemical. We find a group of physiologists, among whom Weber, Ludwig, Du Bois-Reymond, and Helmholtz were noteworthy leaders, devoted to the investigations of physiological facts through the application of measurements and records made by machinery. With these men came into use the time-markers, the myographs, and the ingenious methods of recording blood-pressure, changes in respiration, the responses of muscle and nerve to various forms of stimulation, the rate of transmission of nerve-currents, etc.

The investigation of vital activities by means of measurements and instrumental records has come to represent one especial phase of modern physiology. As might have been predicted, the discoveries and extensions of knowledge resulting from this kind of experimentation have been remarkable, since it is obvious that permanent records made by mechanical devices will rule out many errors; and, moreover, they afford an opportunity to study at leisure phenomena that occupy a very brief time.

The other marked line of physiological investigation has been in the domain of chemistry, where Wöhler, Liebig, Kühne, and others have, through the study of the chemical changes occurring in its body, observed the various activities that take place within the organism. They have reduced all tissues and all parts of the body to chemical analysis, studied the chemical changes in digestion, in respiration, etc. The more recent observers have also made a particular feature of the study of the chemical changes going on within the living matter.

The union of these two chief tendencies into the physico-chemical aspects of physiology has established the modern way of looking upon vital activities. These vital activities are now regarded as being, in their ultimate analysis, due to physical and chemical changes taking place within the living substratum. All along, this physico-chemical idea has been in contest with that of a duality between the body and the life that is manifested in it. The vitalists, then, have had many controversies with those who make their interpretations along physico-chemical lines. We will recollect that vitalism in the hands of the immediate successors of Haller became not only highly speculative, but highly mystical, tending to obscure any close analysis of vital activity and throwing explanations all back into the domain of mysticism. Johannes Müller was also a vitalist, but his vitalism was of a more acceptable form. He thought of changes in the body as being due to vitality—to a living force; but he did not deny the possibility of the transformation of this vital energy into other forms of energy; and upon the basis of Müller's work there has been built up the modern conception that there is found in the human body a particular transformation-form of energy, not a mystical vital force that presides over all manifestations of life.

The advances in physiology, beginning with those of William Harvey, have had immense influence not only upon medicine, but upon all biology. We find now the successful and happy union between physiology and morphology in the work which is being so assiduously carried on to-day under the title of experimental morphology.

The great names in physiology since Müller are numerous, and perhaps it is invidious to mention particular ones; but, inasmuch as Ludwig and Du Bois-Reymond have been spoken of, we may associate with them the names of Sir Michael Foster and Burdon-Sanderson, in England; and of Brücke (one of Müller's disciples) and Verworn, in Germany, as modern leaders whose investigations have promoted advance, and whose clear exposition of the facts and the theories of physiology have added much to the dignity of the science.

CHAPTER X

VON BAER AND THE RISE OF EMBRYOLOGY

ANATOMY investigates the arrangement of organic tissues; embryology, or the science of development, shows how they are produced and arranged. There is no more fascinating division of biological study. As Minot says: "Indeed, the stories which embryology has to tell are the most romantic known to us, and the wildest imaginative creations of Scott or Dumas are less startling than the innumerable and almost incredible shifts of rôle and change of character which embryology has to entertain us with in her histories."

Embryology is one of the most important biological sciences in furnishing clues to the past history of animals. Every organism above the very lowest, no matter how complex, begins its existence as a single microscopic cell, and between that simple state and the fully formed condition every gradation of structure is exhibited. Every time an animal is developed these constructive changes are repeated in orderly sequence, and one who studies the series of steps in development is led to recognize that the process of building an animal's body is one of the most wonderful in all nature.

Rudimentary Organs.—But, strangely enough, the course of development in any higher organism is not straightforward, but devious. Instead of organs being produced in the most direct manner, unexpected by-paths are followed, as when all higher animals acquire gill-clefts and many other rudimentary organs not adapted to their condition of life. Most of the rudimentary organs are transitory, and bear testimony, as hereditary survivals, to the line of ancestry. They are clues by means of which phases in the evolution of animal life may be deciphered.

Bearing in mind the continually shifting changes through which animals pass in their embryonic development, one begins to see why the adult structures of animals are so difficult to understand. They are not only complex; they are also greatly modified. The adult condition of any organ or tissue is the last step in a series of gradually acquired modifications, and is, therefore, the farthest departure from that which is ancestral and archetypal. But in the process of formation all the simpler conditions are exhibited. If, therefore, we wish to understand an organ or an animal, we must follow its development, and see it in simpler conditions, before the great modifications have been added.

The tracing of the stages whereby cells merge into tissues, tissues into organs, and determining how the organs by combinations build up the body, is embryology. On account of the extended applications of this subject in biology, and the light which it throws on all structural studies, we shall be justified in giving its history at somewhat greater length than that adopted in treating of other topics.

Five Historical Periods.—The story of the rise of this interesting department of biology can, for convenience, be divided into five periods, each marked by an advance in general knowledge. These are: (1) the period of Harvey and Malpighi; (2) the period of Wolff; (3) the period of Von Baer; (4) the period from Von Baer to Balfour; and (5) the period of Balfour, with an indication of present tendencies. Among all the leaders Von Baer stands as a monumental figure at the parting of the ways between the new and the old—the sane thinker, the great observer.

The Period of Harvey and Malpighi

In General.—The usual account of the rise of embryology is derived from German writers. But there is reason to depart from their traditions, in which Wolff is heralded as its founder, and the one central figure prior to Pander and Von Baer.

The embryological work of Wolff's great predecessors, Harvey and Malpighi, has been passed over too lightly. Although these men have received ample recognition in closely related fields of investigation, their insight into those mysterious events that culminate in the formation of a new animal has been rarely appreciated. Now and then a few writers, as Brooks and Whitman, have pointed out the great worth of Harvey's work in embryology, but fewer have spoken for Malpighi in this connection. Koelliker, it is true, in his address at the unveiling of the statue of Malpighi, in his native town of Crevalcuore, in 1894, gives him well-merited recognition as the founder of embryology, and the late Sir Michael Foster has written in a similar vein in his delightful *Lectures on the History of Physiology.*

However great was Harvey's work in embryology, I venture to say that Malpighi's was greater when considered as a piece of observation. Harvey's work is more philosophical; he discusses the nature of development, and shows unusual powers as an accurate reasoner. But that part of his treatise devoted to observation is far less extensive and exact than Malpighi's, and throughout his lengthy discussions he has the flavor of the ancients.

Malpighi's work, on the contrary, flavors more of the moderns. In terse descriptions, and with many sketches, he shows the changes in the hen's egg from the close of the first day of development onward.

It is a noteworthy fact that, at the period in which he

lived, Malpighi could so successfully curb the tendency to indulge in wordy disquisitions, and that he was satisfied to observe carefully, and tell his story in a simple way. This quality of mind is rare. As Emerson has said: "I am impressed with the fact that the greatest thing a human soul ever does in this world is to see something, and tell what it saw in a plain way. Hundreds of people can talk for one who can think, but thousands can think for one who can see. To see clearly is poetry, philosophy, and religion all in one." But "to see" here means, of course, to interpret as well as to observe.

Although there were observers in the field of embryology before Harvey, little of substantial value had been produced. The earliest attempts were vague and uncritical, embracing only fragmentary views of the more obvious features of body-formation. Nor, indeed, should we look for much advance in the field of embryology even in Harvey's time. The reason for this will become obvious when we remember that the renewal of independent observation had just been brought about in the preceding century by Vesalius, and that Harvey himself was one of the pioneers in the intellectual awakening. Studies on the development of the body are specialized, involving observations on minute structures and recondite processes, and must, therefore, wait upon considerable advances in anatomy and physiology. Accordingly, the science of embryology was of late development.

Harvey.—Harvey's was the first attempt to make a eritical analysis of the process of development, and that he did not attain more was not owing to limitations of his powers of discernment, but to the necessity of building on the general level of the science of his time, and, further, to his lack of instruments of observation and technique. Nevertheless, Harvey may be considered as having made the first independent advance in embryology.

By clearly teaching, on the basis of his own observations, the gradual formation of the body by aggregation of its parts, he anticipated Wolff. This doctrine came to be known under the title of "epigenesis," but Harvey's epigenesis* was not, as Wolff's was, directed against a theory of pre-delineation of the parts of the embryo, but against the ideas of the medical men of the time regarding the metamorphosis of germinal elements. It lacked, therefore, the dramatic setting which surrounded the work of Wolff in the next century. Had the doctrine of pre-formation been current in Harvev's time, we are quite justified in assuming that he would have assailed it as vigorously as did Wolff.

His Treatise on Generation.—Harvey's embryological work was published in 1651 under the title *Exercitationes de Generatione Animalium*. It embraces not only observations on the development of the chick, but also on the deer and some other mammals. As he was the court physician of Charles I, that sovereign had many deer killed in the park, at intervals, in order to give Harvey the opportunity to study their development.

As fruits of his observation on the chick, he showed the position in which the embryo arises within the egg, *viz.*, in the white opaque spot or cicatricula; and he also corrected Aristotle, Fabricius, and his other predecessors in many particulars.

Harvey's greatest predecessor in this field, Fabricius, was also his teacher. When, in search of the best training in medicine, Harvev took his way from England to Italy, as already recounted, he came under the instruction of Fabricius in Padua. In 1600, Fabricius published sketches showing the development of animals; and, again, in 1625, six years after his death, appeared his illustrated treatise on

* As Whitman has pointed out, Aristotle taught epigenesis as clearly as Harvey, and is, therefore, to be regarded as the founder of that conception.

the development of the chick. Except the figures of Coiter (1573), those of Fabricius were the earliest published illus trations of the kind. Altogether his figures show develop mental stages of the cow, sheep, pig, galeus, serpent, rat, and chick.

Harvey's own treatise was not illustrated. With that singular independence of mind for which he was conspicuous, the vision of the pupil was not hampered by the authority of his teacher, and, trusting only to his own sure observation and reason, he described the stages of development as he saw them in the egg, and placed his own construction on the facts.

One of the earliest activities to arrest his attention in the chick was a pulsating point, the heart, and, from this observation, he supposed that the heart and the blood were the first formations. He says: "But as soon as the egg, under the influence of the gentle warmth of the incubating hen, or of warmth derived from another source, begins to pullulate, this spot forthwith dilates, and expands like the pupil of the eye; and from thence, as the grand center of the egg, the latent plastic force breaks forth and germinates. This first commencement of the chick, however, so far as I am aware, has not yet been observed by any one."

It is to be understood, however, that the descriptive part of his treatise is relatively brief (about 40 pages out of 350 in Willis's translation), and that the bulk of the 106 "exercises" into which his work is divided is devoted to comments on the older writers and to discussions of the nature of the process of development.

The aphorism, "*omne vivum ex ovo*," though not invented by Harvey, was brought into general use through his writings. As used in his day, however, it did not have its full modern significance. With Harvey it meant simply that the embryos of all animals, the viviparous as well as the oviparous, orig-

FIG. 60.—Frontispiece to Harvey's *Generatione Animalium* (1651).

nate in eggs, and it was directed against certain contrary medical theories of the time.

The first edition of his *Generatione Animalium*, London, 1651, is provided with an allegorical frontispiece embodying this idea. As shown in Fig. 60, it represents Jove on a pedestal, uncovering a round box, or ovum, bearing the inscription "*ex ovo omnia*," and from the box issue all forms of living creatures, including also man.

Malpighi.—The observer in embryology who looms into prominence between Harvey and Wolff is Malpighi. He supplied what was greatly needed at the time—an illustrated account of the actual stages in the development of the chick from the end of the first day to hatching, shorn of verbose references and speculations.

His observations on development are in two separate memoirs, both sent to the Royal Society in 1672, and pub lished by the Society in Latin, under the titles *De Formatione Pulli in Ovo* and *De Ovo Incubato*. The two taken together are illustrated by twelve plates containing eighty-six figures, and the twenty-two quarto pages of text are nearly all devoted to descriptions, a marked contrast to the 350 pages of Harvey unprovided with illustrations.

His pictures, although not correct in all particulars, represent what he was able to see, and are very remarkable for the age in which they were made, and considering the instruments of observation at his command. They show successive stages from the time the embryo is first outlined, and, taken in their entirety, they cover a wide range of stages.

His observations on the development of the heart, comprising twenty figures, are the most complete. He clearly illustrates the aortic arches, those transitory structures of such great interest as showing a phase in ancestral history.

He was also the first to show by pictures the formation of the head-fold and the neural groove, as well as the brain-

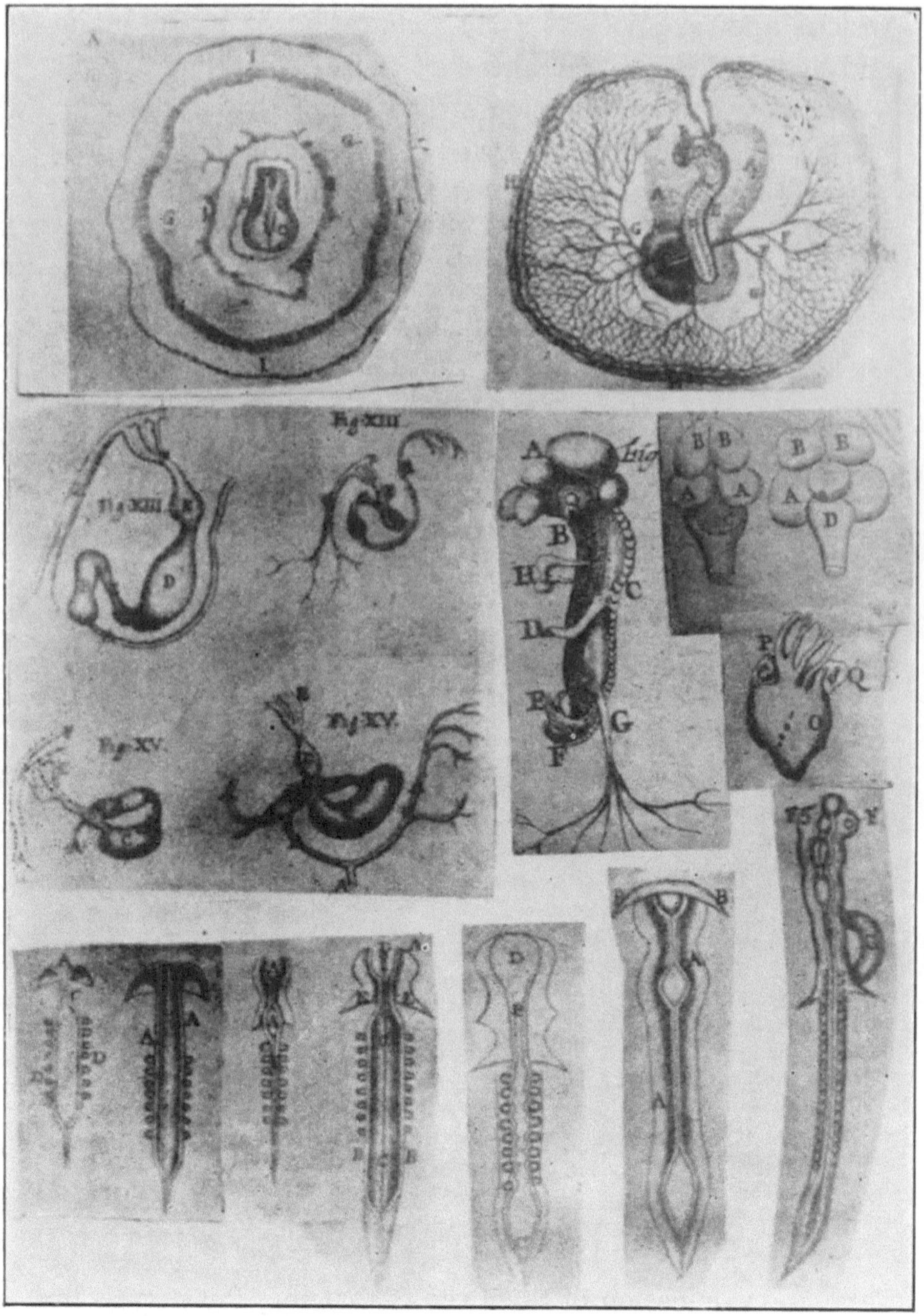

FIG. 61.—Selected Sketches from Malpighi's Works, Showing Stages in the Development of the Chick (1672).

vesicles and eye-pockets. His delineation of heart, brain, and eye-vesicles are far ahead of those illustrating Wolff's *Theoria Generationis*, made nearly a hundred years later.

Fig. 61 shows a few selected sketches from the various plates of his embryological treatises, to compare with those of Wolff. (See Fig. 63.)

The original drawings for *De Ovo Incubato*, still in possession of the Roval Society, are made in pencil and red chalk,

FIG. 62.—MARCELLO MALPIGHI, 1628–1694.

and an examination of them shows that they far surpass the reproductions in finish and accuracy.

While Harvey taught the gradual formation of parts, Malpighi, from his own observations, supposed the rudiments of the embryo to pre-exist within the egg. He thought that, possibly, the blood-vessels were in the form of tubes, closely wrapped together, which by becoming filled with blood were distended. Nevertheless, in the treatises mentioned above he is very temperate in his expressions on the whole matter, and evidently believed in the new formation of many parts.

The portrait of Malpighi shown in Fig. 62 is taken from his life by Atti. From descriptions of his personal appearance (see page 58) one would think that this is probably a better likeness than the strikingly handsome portrait painted by Tabor, and presented by Malpighi to the Royal Society of London. For a reproduction of the latter see page 59.

Malpighi's Rank.—On the whole, Malpighi should rank above Harvey as an embryologist, on account of his discoveries and fuller representation, by drawings and descriptions, of the process of development. As Sir Michael Foster has said: "The first adequate description of the long series of changes by which, as they melt the one into the other, like dissolving views, the little white opaque spot in the egg is transformed into the feathered, living, active bird, was given by Malpighi. And where he left it, so for the most part the matter remained until even the present century. For this reason we may speak of him as the founder of embryology."

The Period of Wolff

Between Harvey and Wolff, embryology had become dominated by the theory that the embryo exists already pre-formed within the egg, and, as a result of the rise of this new doctrine, the publications of Wolff had a different setting from that of any of his predecessors. It is only fair to say that to this circumstance is owing, in large part, the prominence of his name in connection with the theory of epigenesis.

As we have already seen, Harvey, more than a century before the publications of Wolff, had clearly taught that development is a process of gradual becoming. Nevertheless, Wolff's work, as opposed to the new theory, was very important.

While the facts fail to support the contention that he was the founder of epigenesis, it is to be remembered that he has claims in other directions to rank as the foremost student of embryology prior to Von Baer.

As a preliminary to discussing Wolff's position, we should bring under consideration the doctrine of pre-formation and encasement.

Rise of the Theory of Pre-delineation.—The idea of pre-formation in its first form is easily set forth. Just as when we examine a seed we find within an embryo plantlet, so it was supposed that the various forms of animal life existed in miniature within the egg. The process of development was supposed to consist of the expansion or unfolding of this pre-formed embryo. The process was commonly illustrated by reference to flower-buds. "Just as already in a small bud all the parts of the flower, such as stamens and colored petals, are enveloped by the green and still undeveloped sepals; just as the parts grow in concealment and then suddenly expand into a blossom, so also in the development of animals, it was thought that the already present, small but transparent parts grow, gradually expand, and become discernible." (Hertwig.) From the feature of unfolding this was called in the eighteenth centurv the theory of *evolution*, giving to that term quite a different meaning from that attached to it at the present time.

This theory, strange as it may seem to us now, was founded on a basis of actual observation—not entirely on speculation. Although it was a product of the seventeenth century, from several printed accounts one is likely to gather the impression that it arose in the eighteenth century, and that

Bonnet, Haller, and Leibnitz were among its founders. This implication is in part fostered by the circumstance that Swammerdam's *Biblia Naturæ*, which contains the germ of the theory, was not published until 1737—more than half a century after his death—although the observations for it were completed before Malpighi's first paper on embryology was published in 1672. While it is well to bear in mind that date of publication, rather than date of observation, is accepted as establishing the period of emergence of ideas, there were other men, as Malpighi and Leeuwenhoek, contemporaries of Swammerdam, who published in the seventeenth century the basis for this theory

Malpighi supposed (1672) the rudiment of the embryo to pre-exist within the hen's egg, because he observed evidences of organization in the unincubated egg. This was in the heat of the Italian summer (in July and August, as he himself records), and Dareste suggests that the developmental changes had gone forward to a considerable degree before Malpighi opened the eggs. Be this as it may, the imperfection of his instruments and technique would have made it very difficult to see anything definitely in stages under twenty-four hours.

In reference to his observations, he says that in the unincubated egg he saw a small embryo enclosed in a sac which he subjected to the rays of the sun. "Frequently I opened the sac with the point of a needle, so that the animals contained within might be brought to the light, nevertheless to no purpose; for the individuals were so jelly-like and so very small that they were lacerated by a light stroke. Therefore, it is right to confess that the beginnings of the chick pre-exist in the egg, and have reached a higher development in no other way than in the eggs of plants." ("Quare *pulli stamina* in ovo *præexistere*, altiorémque originem nacta esse fateri convenit, haud dispari ritu, ac in Plantarum ovis.")

Swammerdam (1637–1680) supplied a somewhat better basis. He observed that the parts of the butterfly, and other insects as well, are discernible in the chrysalis stage. Also, on observing caterpillars just before going into the pupa condition, he saw in outline the organs of the future stage, and very naturally concluded that development consists of an expansion of already formed parts.

A new feature was introduced through the discovery, by Leeuwenhoek, about 1677,* of the fertilizing filaments of eggs. Soon after, controversies began to arise as to whether the embryo pre-existed in the sperm or in the egg. By Leeuwenhoek, Hartsoeker, and others the egg was looked upon as simply a *nidus* within which the sperm developed, and they asserted that the future animal existed in miniature in the sperm. These controversies gave rise to the schools of the animalculists, who believed the sperm to be the animal germ, and of the ovulists, who contended for the ovum in that rôle.

It is interesting to follow the metaphysical speculations which led to another aspect of the doctrine of pre-formation. There were those, notably Swammerdam, Leibnitz, and Bonnet, who did not hesitate to follow the idea to the logical consequence that, if the animal germ exists pre-formed, one generation after another must be encased within it. This gave rise to the fanciful idea of encasement or *emboîtement*, which was so greatly elaborated by Bonnet and, by Leibnitz, applied to the development of the soul. Even Swammerdam (who, by the way, though a masterly observer, was always a poor generalizer) conceived of the germs of all forthcoming generations as having been located in the common mother Eve, all closely encased one within the other, like the boxes of a Japanese juggler. The end of the human race was con-

* The discovery is also attributed to Hamm, a medical student, and to Hartsoeker, who claimed priority in the discovery.

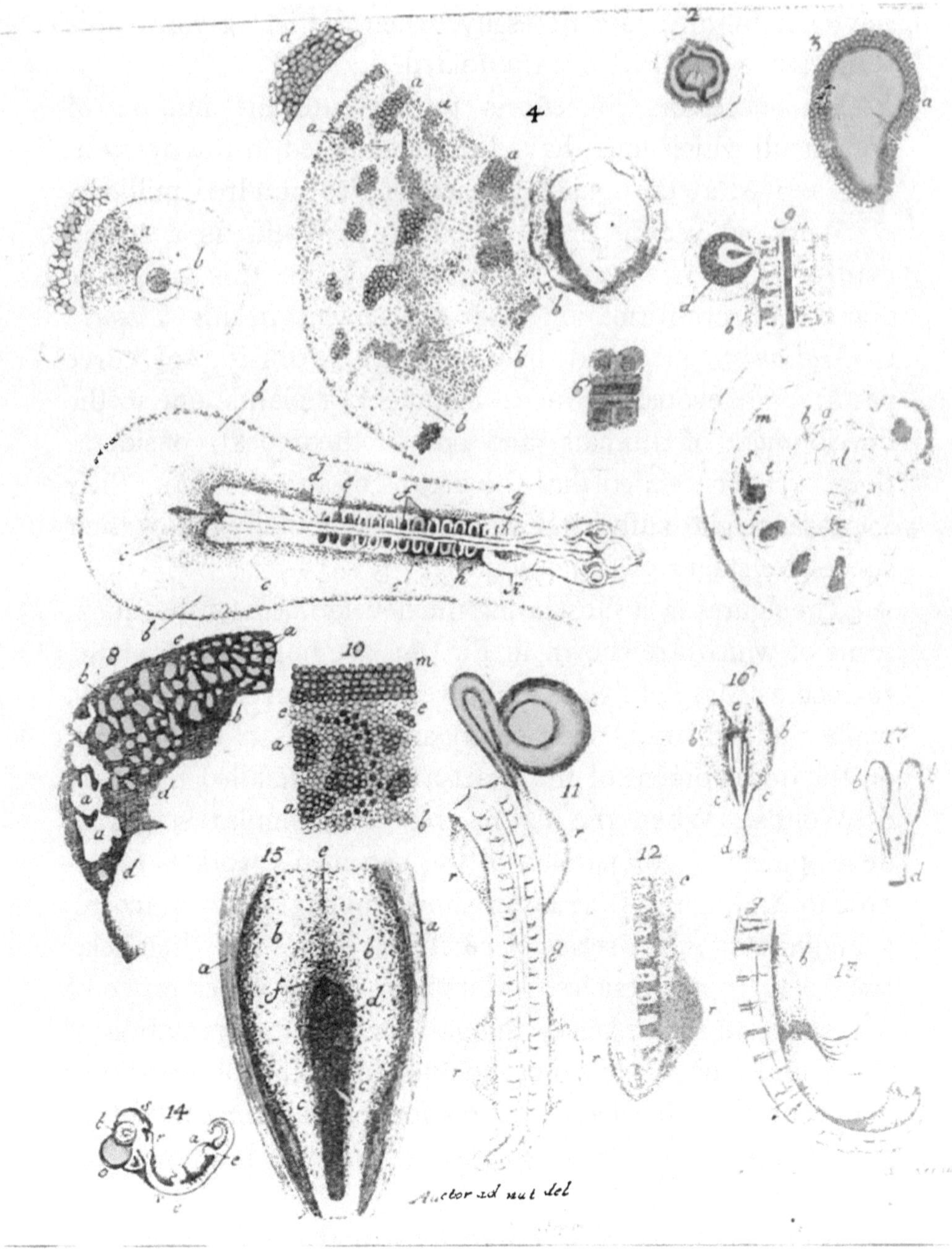

Fig. 63.—Plate from Wolff's *Theoria Generationis* (1759), Showing Stages in the Development of the Chick.

ceived of by him as a necessity, when the last germ of this wonderful series had been unfolded.

His successors, in efforts to compute the number of homunculi which must have been condensed in the ovary of Eve, arrived at the amazing result of two hundred millions.

Work of Wolff.—Friedrich Kaspar Wolff, as a young man of twenty-six years, set himself against this grotesque doctrine of pre-formation and encasement in his *Theoria Generationis*, published in 1759. This consists of three parts: one devoted to the development of plants, one to the development of animals, and one to theoretical considerations. He contended that the organs of animals make their appearance gradually, and that he could actually follow their successive stages of formation.

The figures in it illustrating the development of the chick, some of which are shown in Fig. 63, are not, on the whole, so good as Malpighi's. Wolff gives, in all, seventeen figures, while Malpighi published eighty-six, and his twenty figures on the development of the heart are more detailed than any of Wolff's. When the figures represent similar stages of development, a comparison of the two men's work is favorable to Malpighi. The latter shows much better, in corresponding stages, the series of cerebral vesicles and their relation to the optic vesicles. Moreover, in the wider range of his work, he shows many things—such as the formation of the neural groove, etc.—not included in Wolff's observations. Wolff, on the other hand, figures for the first time the primitive kidneys, or "Wolffian bodies," of which he was the discoverer.

Although Wolff was able to show that development consists of a gradual formation of parts, his theory of development was entirely mystical and unsatisfactory. The fruitful idea of germinal continuity had not yet emerged, and the thought that the egg has inherited an organization from the past was yet to be expressed. Wolff was, therefore, in the same quandary as his predecessors when he undertook to explain development. Since he assumed a total lack of organization in the beginning, he was obliged to make development "miraculous" through the action on the egg of a hyperphysical agent. From a total lack of organization, he conceived of its being lifted to the highly organized product through the action of a "*vis essentialis corporis.*"

He returned to the problem of development later, and, in 1768–1769, published his best work in this field on the development of the intestine * This is a very original and strong piece of observational work. While his investigations for the *Theoria Generationis* did not reach the level of Malpighi's, those of the paper of 1768 surpassed them and held the position of the best piece of embryological work up to that of Pander and Von Baer. This work was so highly appreciated by Von Baer that he said: "It is the greatest masterpiece of scientific observation which we possess." In it he clearly demonstrated that the development of the intestine and its appendages is a true process of becoming. Still later, in 1789, he published further theoretical considerations.

Opposition to Wolff's Views.—But all Wolff's work was launched into an uncongenial atmosphere. The great physiologist Haller could not accept the idea of epigenesis, but opposed it energetically, and so great was his authority that the views of Wolff gained no currency. This retarded progress in the science of animal development for more than a half-century.

Bonnet was also a prolific writer in opposition to the ideas of Wolff, and we should perhaps have a portrait of him (Fig. 64) as one of the philosophical naturalists of the time. His prominent connection with the theory of pre-delineation

* *De Formatione Intestinorum, Nova Commentar, Ac. Sci. Petrop.*, St. Petersburg, XII., 1768; XIII., 1769.

in its less grotesque form, his discoverv of the development of the eggs of plant-lice without previous fertilization, his researches on regeneration of parts in polyps and worms, and other observations place him among the conspicuous

FIG. 64.—CHARLES BONNET, 1720–1793.

naturalists of the period. His system of philosophy, which has been carefully analyzed by Whitman, is designated by that writer as a system of negations.

In 1821, J. Fr. Meckel, recognizing the great value of

Wolff's researches on the development of the intestines, rescued the work from neglect and obscurity by publishing a German translation of the same, and bringing it to the attention of scholars. From that time onward Wolff's labor was fruitful.

His *De Formatione Intestinorum* rather than his *Theoria Generationis* embodies his greatest contribution to embryology. Not only is it a more fitting model of observation, but in it he foreshadows the idea of germ-layers in the embryo, which, under Pander and Von Baer, became the fundamental conception in structural embryology. Throughout his researches both early and late, he likens the embryonic rudiments, which precede the formation of organs, to leaflets. In his work of 1768 he described in detail how the leaf-like layers give rise to the systems of organs; showing that the nervous system arises first from a leaf-like layer, and is followed, successively, by a flesh layer, the vascular system, and lastly, by the intestinal canal—all arising from original leaf-like layers.

In these important generalizations, although they are verbally incorrect, he reached the truth as nearly as it was possible at the time, and laid the foundation of the germ-layer theory.

Wolff was a man of great power as an observer, and although his influence was for a long time retarded, he should be recognized as the foremost investigator in embryology before Von Baer.

Few Biographical Facts.—The little known of his life is gained through his correspondence and a letter by his amanuensis. Through personal neglect, and hostility to his work, he could not secure a foothold in the universities of Germany, and, in 1764, on the invitation of Catherine of Russia, he went to the Academy of Sciences at St. Petersburg, where he spent the last thirty years of his life.

It has been impossible to discover a portrait of Wolff, although I have sought one in various ways for several years. The secretary of the Academy of Sciences at St. Petersburg writes that no portrait of Wolff exists there, and that the Academy will gratefully receive information from any source regarding the existence of a portrait of the great academician.

His sincere and generous spirit is shown in his correspondence with Haller, his great opponent. "And as to the matter of contention between us, I think thus: For me, no more than for you, glorious man, is truth of the very greatest concern. Whether it chance that organic bodies emerge from an invisible into a visible condition, or form themselves out of the air, there is no reason why I should wish the one were truer than the other, or wish the one and not the other. And this is your view also, glorious man. We are investigating for truth only; we seek that which is true. Why then should I contend with you?" (Quoted from Wheeler.)

The Period of Von Baer

What Johannes Müller was for physiology, von Baer was for embryology; all subsequent growth was influenced by his investigations.

The greatest classic in embryology is his *Development of Animals* (*Entwickelungsgeschichte der Tiere—Beobachtung und Reflexion*), the first part of which was published in 1828, and the work on the second part completed in 1834, although it was not published till 1837. This second part was never finished according to the plan of Von Baer, but was issued by his publisher, after vainly waiting for the finished manuscript. The final portion, which Von Baer had withheld, in order to perfect in some particulars, was published in 1888, after his death, but in the form in which he left it in 1834.

The observations for the first part began in 1819, after he had received a copy of Pander's researches, and covered a period of seven years of close devotion to the subject; and the observations for the last part were carried on at intervals for several years.

It is significant of the character of his *Reflexionen* that, although published before the announcement of the cell-theory, and before the acceptance of the doctrine of organic evolution, they have exerted a molding influence upon embryology to the present time. The position of von Baer in embryology is owing as much to his sagacity in speculation as to his powers as an observer. "Never again have observation and thought been so successfully combined in embryological work" (Minot).

Von Baer was born in 1792, and lived on to 1876, but his enduring fame in embryology rests on work completed more than forty years before the end of his useful life. After his removal from Königsberg to St. Petersburg, in 1834, he very largely devoted himself to anthropology in its widest sense, and thereby extended his scientific reputation into other fields.

If space permitted, it would be interesting to give the biography* of this extraordinary man, but here it will be necessary to content ourselves with an examination of his portraits and a brief account of his work.

Portraits.—Several portraits of von Baer showing him at different periods of his life have been published. A very attractive one, taken in his early manhood, appeared in *Harper's Magazine* for 1898. The expression of the face is poetical, and the picture is interesting to compare with the more matured, sage-like countenance forming the frontispiece

* Besides biographical sketches by Stieda, Waldeyer, and others, we have a very entertaining autobiography of Von Baer, published in 1864, for private circulation, but afterward (1866) reprinted and placed on sale.

of Stieda's *Life of Von Baer* (see Fig. 65). This, perhaps the best of all his portraits, shows him in the full development of his powers. An examination of it impresses one

FIG. 65.—KARL ERNST VON BAER, 1792–1876.

with confidence in his balanced judgment and the thoroughness and profundity of his mental operations.

The portrait of Von Baer at about seventy years of age,

reproduced in Fig. 66, is, however, destined to be the one by which he is commonly known to embryologists, since it forms the frontispiece of the great cooperative *Handbook of Em-*

FIG. 66.—VON BAER AT ABOUT SEVENTY YEARS OF AGE.

bryology just published under the editorship of Oskar Hertwig.

Von Baer's Especial Service.—Apart from special discoveries, Von Baer greatly enriched embryology in three directions: In the first place, he set a higher standard for all work in embryology, and thereby lifted the entire science to a higher level. Activity in a great field of this kind is, with the rank and file of workers, so largely imitative that this feature of his influence should not be overlooked. In the second place, he established the germ-layer theory, and, in the third, he made embryology comparative.

In reference to the germ-layer theory, it should be recalled that Wolff had distinctly foreshadowed the idea by showing that the material out of which the embryo is constructed is, in an early stage of development, arranged in the form of leaf-like layers. He showed specifically that the alimentary canal is produced by one of these sheet-like expansions folding and rolling together.

Pander, by observations on the chick (1817), had extended the knowledge of these layers and elaborated the conception of Wolff. He recognized the presence of three primary layers—an outer, a middle, and an inner—out of which the tissues of the body are formed.

The Germ-Layers.—But it remained for Von Baer,* by extending his observations into all the principal groups of animals, to raise this conception to the rank of a general law of development. He was able to show that in all animals

* It is of more than passing interest to remember that Pander and Von Baer were associated as friends and fellow-students, under Döllinger at Würzburg. It was partly through the influence of Von Baer that Pander came to study with Döllinger, and took up investigations on development. His ample private means made it possible for him to bear the expenses connected with the investigation, and to secure the services of a fine artist for making the illustrations. The result was a magnificently illustrated treatise. His unillustrated thesis in Latin (1817) is more commonly known, but the illustrated treatise in German is rarer. Von Baer did not take up his researches seriously until Pander's were published. It is significant of their continued harmonious relations that Von Baer's work is dedicated "An meinen Jugendfreund, Dr. Christian Pander."

except the very lowest there arise in the course of development leaf-like layers, which become converted into the "fundamental organs" of the body.

Now, these elementary layers are not definitive tissues of the body, but are embryonic, and therefore may appropriately be designated "germ-layers." The conception that these germ-layers are essentially similar in origin and fate in all animals was a fuller and later development of the germ-layer theory, a conception which dominated embryological study until a recent date.

Von Baer recognized four such layers; the outer and inner ones being formed first, and subsequently budding off a middle layer composed of two sheets. A little later (1845) Remak recognized the double middle layer of Von Baer as a unit, and thus arrived at the fundamental conception of three layers—the ecto-, endo-, and mesoderm—which has so long held sway. For a long time after Von Baer the aim of embryologists was to trace the history of these germ-layers, and so in a wider and much qualified sense it is to-day.

It will ever stand to his credit, as a great achievement, that Von Baer was able to make a very complicated feature of development clear and relatively simple. Given a leaf-like rudiment, with the layers held out by the yolk, as is the case in the hen's egg, it was no easy matter to conceive how they are transformed into the nervous system, the body-wall, the alimentary canal, and other parts, but Von Baer saw deeply and clearly that the fundamental anatomical features of the body are assumed by the leaf-like rudiments being rolled into tubes.

Fig. 67 shows four sketches taken from the plates illustrating von Baer's work. At *A* is shown a stage in the formation of the embryonic envelope, or amnion, which surrounds the embryos of all animals above the class of amphibia. *B*, another figure of an ideal section, shows that, long before the day of microtomes, Von Baer made use of sections to represent the relationships of his four germ-layers. At *C* and *D* is represented diagrammatically the way in which these layers are rolled into tubes. He showed that the central nervous system arose in the form of a tube, from the outer layer; the body-wall in the form of a tube, composed of skin and muscle layers; and the alimentary tube from mucous and vascular layers.

The generalization that embrvos in development tend to recapitulate their ancestral history is frequently attributed to Von Baer, but the qualified way in which he suggests something of the sort will not justify one in attaching this conelusion to his work.

Von Baer was the first to make embryology truly comparative, and to point out its great value in anatomy and zoölogy. By embryological studies he recognized four types of organization—as Cuvier had done from the standpoint of comparative anatomy. But, since these types of organiza tion have been greatly changed and subdivided, the importance of the distinction has faded away. As a distinct break, however, with the old idea of a linear scale of being it was of moment.

Among his especially noteworthy discoveries may be mentioned that of the egg of mammals (1827), and the notochord as occurring in all vertebrate animals. His discovery of the mammalian egg had been preceded by Purkinje's observations upon the germinative spot in the bird's egg (1825).

Von Baer's Rank.—Von Baer has come to be dignified with the title of the "father of modern embryology." No man could have done more in his period, and it is owing to his superb intellect, and to his talents as an observer, that he accomplished what he did. As Minot says: "He worked out, almost as fully as was possible at this time, the genesis

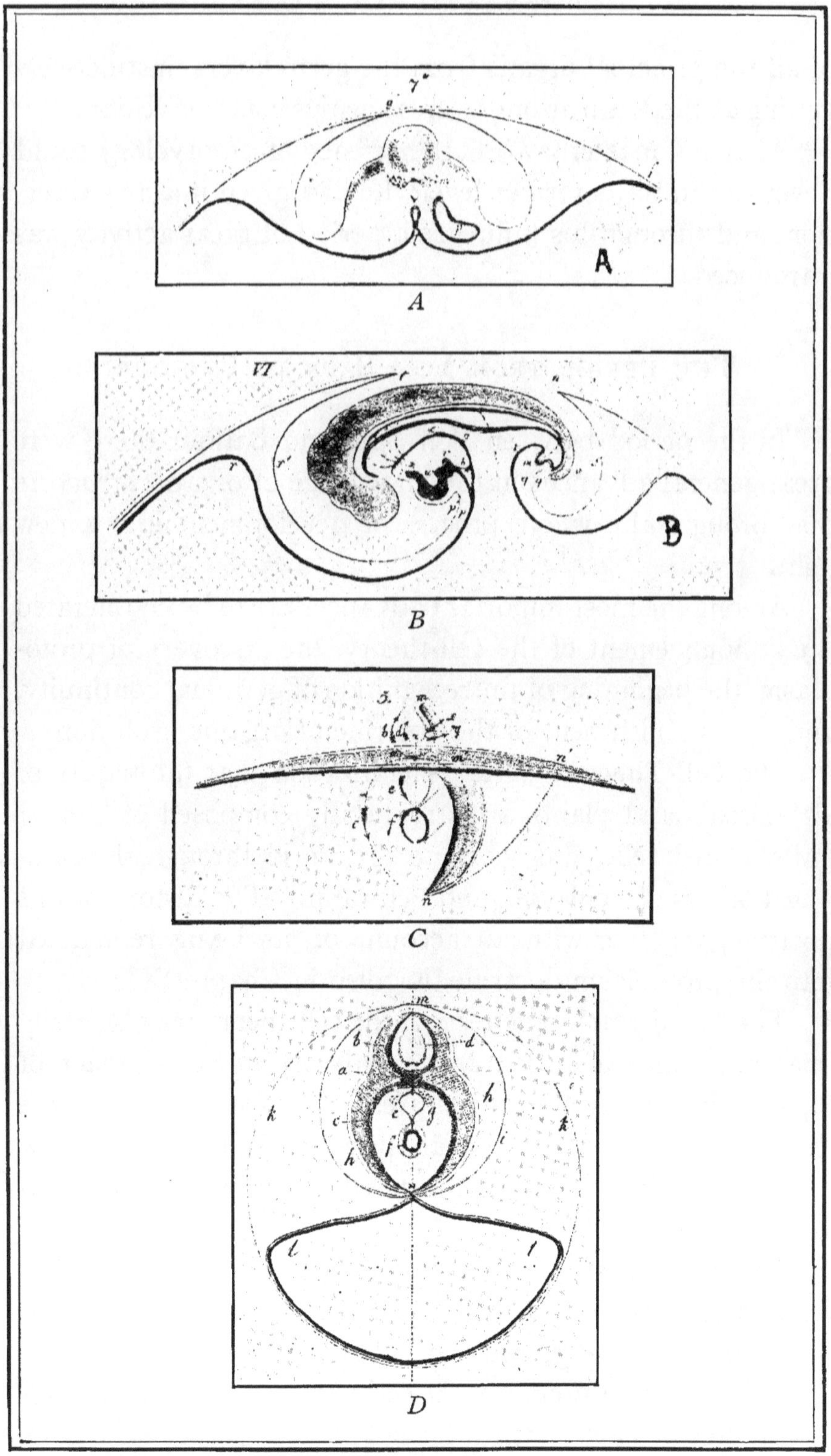

FIG. 67.—Sketches from Von Baer's Embryological Treatise (1828).

of all the principal organs from the germ-layers, instinctively getting at the truth as only a great genius could have done."

After his masterly work, the science of embryology could never return to its former level; he had given it a new direction, and through his influence a period of great activity was introduced.

The Period from Von Baer to Balfour

In the period between Von Baer and Balfour there were great general advances in the knowledge of organic structure that brought the whole process of development into a new light.

Among the most important advances are to be enumerated the announcement of the cell-theory, the discovery of protoplasm, the beginning of the recognition of germinal continuity, and the establishment of the doctrine of organic evolution.

The Cell-Theory.—The generalization that the tissues of all animals and plants are structurally composed of similar units, called cells, was given to the world through the combined labors of Schleiden and Schwann. The history of this doctrine, together with an account of its being remodeled into the protoplasm doctrine, is given in Chapter XII.

The broad-reaching effects of the cell-theory may be easily imagined, since it united all animals on the broad place of likeness in microscopic structure. Now for the first time the tissues of the body were analyzed into their units; now for the first time was comprehended the nature of the germ-layers of Von Baer.

Among the first questions to emerge in the light of the new researches were concerning the origin of cells in the organs the tissues, and the germ-layers. The road to the investigation of these questions was already opened, and it was followed, step by step, until the egg and the sperm came to be recognized as modified cells. This position was reached, for the egg, about 1861, when Gegenbaur showed that the eggs of all vertebrate animals, regardless of size and condition, are in reality single cells. The sperm was put in the same category about 1865.

The rest was relatively easy: the egg, a single cell, by successive divisions produces many cells, and the arrangement of these into primary embryonic layers brings us to the starting-point of Wolff and Von Baer. The cells, continuing to multiply by division, not only increase in number, but also undergo changes through division of physiological labor, whereby certain groups are set apart to perform a particular part of the work of the body. In this way arise the various tissues of the body, which are, in reality, similar cells performing a similar function. Finally, from combinations of tissues, the organs are formed.

But the egg, before entering on the process of develop ment, must be stimulated by the union of the sperm with the nucleus of the egg, and thus the starting-point of every animal and plant, above the lowest group, proves to be a single cell with protoplasm derived from two parents. While questions regarding the origin of cells in the body were being answered, the foundation for the embryological study of heredity was also laid.

Advances were now more rapid and more sure; flashes of morphological insight began to illuminate the way, and the facts of isolated observations began to fit into a harmonized whole.

Apart from the general advances of this period, mentioned in other connections, the work of a few individuals requires notice.

Rathke and Remak were engaged with the broader aspects of embryology, as well as with special investigations. From Rathke's researches came great advances in the knowledge of the development of insects and other invertebrates, and Remak is notable for similar work with the vertebrates. As already mentioned, he was the first to recognize the middle layer as a unit, through which the three germ-layers of later embryologists emerged into the literature of the subject.

Koelliker, 1817–1905, the veteran embryologist, for so many years a professor in the University of Würzburg, carried on investigations on the segmentation of the egg. Besides work on the invertebrates, later he followed with care the development of the chick and the rabbit; he encompassed the whole field of embryology, and published, in 1861 and again in 1876, a general treatise on vertebrate embryology, of high merit. The portrait of this distinguished man is shown in Chapter VIII, where also his services as a histologist are recorded.

Huxley took a great step toward unifying the idea of germ-layers throughout the animal kingdom, when he maintained in 1849, that the two cell-layers in animals like the hydra and oceanic hydrozoa correspond to the ectoderm and endoderm of higher animals.

Kowalevsky (Fig. 68) made interesting discoveries of a general bearing. In 1866 he showed the practical identity, in the early stages of development, between one of the lowest vertebrates (amphioxus) and a tunicate. The latter up to that time had been considered an invertebrate, and the effect of Kowalevsky's observations was to break down the sharply limited line supposed to exist between the invertebrates and the vertebrates. This was of great influence in subsequent work. Kowalevsky also founded the generalization that all animals in development pass through a gastrula stage—a doctrine associated, since 1874, with the name of Haeckel under the title of the gastræa theory.

Beginning of the Doctrine of Germinal Continuity. The conception that there is unbroken continuity of germinal substance between all living organisms, and that the egg and the sperm are endowed with an inherited organization of great complexity, has become the basis for all current theories of heredity and development. So much is involved in this conception that, in the present decade, it has been designated (Whitman) "the central fact of modern biology." The first clear expression of it is found in Virchow's *Cellular Pathology*, published in 1858. It was not, however, until the

FIG. 68.—A. KOWALEVSKY, 1840–1901.

period of Balfour, and through the work of Fol, Van Beneden (chromosomes, 1883), Boveri, Hertwig, and others, that the great importance of this conception began to be appreciated, and came to be woven into the fundamental ideas of development.

Influence of the Doctrine of **Organic Evolution.**—This doctrine, although founded in its modern sense by Lamarck in the early part of the nineteenth century, lay dormant until Darwin, in 1859, brought a new feature into its discussion by emphasizing the factor of natural selection. The general acceptance of the doctrine, which followed after fierce opposition, had, of course, a profound influence on embryology. The latter science is so intimately concerned with the genealogy of animals and plants, that the newly accepted doctrine, as affording an explanation of this genealogy, was the thing most needed.

The development of organisms was now seen in the light of ancestral history, rudimentary organs began to have meaning as hereditary survivals, and the whole process of development assumed a different aspect. This doctrine supplied a new impulse to the interpretation of nature at large, and of the embryological record in particular. The meaning of the embryological record was so greatly emphasized in the period of Balfour that it will be commented upon under the next division of our subject.

The period between Von Baer and Balfour proved to be one of great importance on account of the general advances in knowledge of all organic nature. Observations were moving toward a better and more consistent conception of the structure of animals and plants. A new comparative anatomy, more profound and richer in meaning than Cuvier's, was arising. The edifice on the foundation of Von Baer's work was now emerging into recognizable outlines.

The Period of Balfour, with an Indication of Present Tendencies

Balfour's Masterly Work.—The workers of this period inherited all the accumulations of previous efforts, and the time was ripe for a new step. Observations on the development of different animals, vertebrates and invertebrates, had accumulated in great number, but they were scattered through technical periodicals, transactions of learned societies, monographs, etc., and there was no compact science of embryology with definite outlines. Balfour reviewed all this mass of information, digested it, and molded it into an organized whole. The results were published in the form of two volumes with the title of *Comparative Embryology.* This book of "almost priceless value" was given to the world in 1880–1881. It was a colossal undertaking, but Balfour was

FIG. 69.—FRANCIS M. BALFOUR, 1851–1882.

a phenomenal worker. Before his untimely death at the age of thirty-one, he had been able to complete this work and to produce, besides, a large number of technical researches. The period of Balfour is taken arbitrarily in this volume as beginning about 1874, when he published, with Michael Foster, *The Elements of Embryology.*

His University Career.—Balfour (Fig. 69) was born in

1851. During his days of preparation for the university he was a good student, but did not exhibit in any marked way the powers for which later he became distinguished. At Cambridge, his distinguished teacher, the late Sir Michael Foster, recognized his great talents, and encouraged him to begin work in embryology. His labors in this field once begun, he threw himself into it with great intensity. He rose rapidly to a professorship in Cambridge, and so great was his enthusiasm and earnestness as a lecturer that in seven years "voluntary attendance on his classes advanced from ten to ninety." He was also a stimulator of research, and at the time of his death there were twenty students engaged in his laboratory on problems of development.

He was distinguished for personal attractiveness, and those who met him were impressed with his great sincerity, as well as his personal charm. He was welcomed as an addition to the select group of distinguished scientific men of England, and a great career was predicted for him. Huxley, when he felt the call, at a great personal sacrifice, to lay aside the more rigorous pursuits of scientific research, and to devote himself to molding science into the lives of the people, said of Balfour: "He is the only man who can carry out my work."

His Tragic Fate.—But that was not destined to be. The story of his tragic end need be only referred to. After completing the prodigious labor on the *Comparative Embryology* he went to Switzerland for recuperation, and met his death, with that of his guide, by slipping from an Alpine height into a chasm. His death occurred in July, 1882.

The memorial edition of his works fills four quarto vol umes, but the "Comparative Embryology" is Balfour's monument, and will give him enduring fame. It is not only a digest of the work of others, but contains also general considerations of a far-seeing quality. He saw developmental processes in the light of the hypothesis of organic evolution. His speculations were sufficiently reserved, and nearly always luminous. It is significant of the character of this work to say that the speculations contained in the papers of the rank and file of embryological workers for more than two decades, and often fondly believed to be novel, were for the most part anticipated by Balfour, and were also better expressed, with better qualifications.

The reading of ancestral history in the stages of development is such a characteristic feature of the embryological work of Balfour's period that some observations concerning it will now be in place.

Interpretation of the Embryological Record.—Perhaps the most impressive feature of animal development is the series of similar changes through which all pass in the embryo. The higher animals, especially, exhibit all stages of organization from the unicellular fertilized ovum to the fully formed animal so far removed from it. The intermediate changes constitute a long record, the possibility of interpreting which has been a stimulus to its careful examination.

Meckel, in 1821, and later Von Baer, indicated the close similarity between embryonic stages of widely different animals; Von Baer, indeed, confessed that he was unable to distinguish positively between a reptile, a bird, and a mammalian embryo in certain early stages of growth.

In addition to this similarity, which is a constant feature of the embryological record, there is another one that may be equally significant; *viz.*, in the course of embryonic history, sets of rudimentary organs arise and disappear. Rudimentary teeth make their appearance in the embryo of the whalebone whale, but they are transitory and soon disappear without having been of service to the animal. In the embryos of all higher vertebrates, as is well known, gill-clefts and gill-arches with an appropriate circulation, make their appearance, but disappear long before birth. These indications, and similar ones, must have some meaning.

Now whatever qualities an animal exhibits after birth are attributed to heredity. May it not be that all the intermediate stages are also inheritances, and, therefore, represent phases in ancestral history? If they be, indeed, clues to ancestral conditions, may we not, by patching together our observations, be able to interpret the record, just as the history of ancient peoples has been made out from fragments in the shape of coins, vases, implements, hieroglyphics, inscriptions, etc.?

The Recapitulation Theory.—The results of reflection in this direction led to the foundation of the *recapitulation theory*, according to which animals are supposed, in their individual development, to recapitulate to a considerable degree phases of their ancestral history. This is one of the widest generalizations of embryology. It was suggested in the writings of Von Baer and Louis Agassiz, but received its first clear and complete expression in 1863, in the writings of Fritz Müller.

Although the course of events in development is a record, it is, at best, only a fragmentary and imperfect one. Many stages have been dropped out, others are unduly prolonged or abbreviated, or appear out of chronological order, and, besides this, some of the structures have arisen from adaptation of a particular organism to its conditions of development, and are, therefore, not ancestral at all, but, as it were, recent additions to the text. The interpretation becomes a difficult task, which requires much balance of judgment and profound analysis.

The recapitulation theory was a dominant note in all Balfour's speculations, and in that of his contemporary and fellow-student Marshall. It has received its most sweeping application in the works of Ernst Haeckel.

Widely spread throughout recent literature is to be noted a reaction against the too wide and unreserved application of this doctrine. This is naturally to be expected, since it is the common tendency in all fields of scholarship to demand

FIG 70 —OSKAR HERTWIG IN 1890.

a more critical estimate of results, and to undergo a reaction from the earlier crude and sweeping conclusions.

Nearly all problems in anatomy and structural zoölogy are approached from the embryological side, and, as a consequence, the work of the great army of anatomists and zoölogists has been in a measure embryological. Many of them have produced beautiful and important researches, but the work is too extended to admit of review in this connection.

Oskar Hertwig, of Berlin (Fig. 70), is one of the representative embryologists of Europe, while, in this country, lights of the first magnitude are Brooks, Minot, Whitman, E. B. Wilson, and others.

Although no attempt is made to review the researches of the recent period, we cannot pass entirely without mention the discovery of chromosomes, and of their reduction in the ripening of the egg and in the formation of sperm. This has thrown a flood of light on the phenomena of fertilization, and has led to the recognition of chromosomes as probably the bearers of heredity. The nature of fertilization, investigated by Fol, O. Hertwig, and others, formed the starting-point for a series of brilliant discoveries.

The embryological investigations of the late Wilhelm His (Fig. 71) are also deserving of especial notice. His luminous researches on the development of the nervous system, the origin of nerve fibers, and his analysis of the development of the human embryo are all very important.

Recent Tendencies. Experimental Embryology.—Soon after the publication of Balfour's great work on "Comparative Embryology," a new tendency in research began to appear which led onward to the establishment of experimental embryology. All previous work in this field had been concerned with the structure, or architecture, of organisms, but now the physiological side began to receive attention. Whitman has stated with great aptness the interdependence of these two lines of work, as follows: "Morphology raises the question, How came the organic mechanism into existence? Has it had a history, reaching its present stage of perfection through a long series of gradations, the first term of which was a relatively simple stage? The embryological history is traced

FIG. 71.—WILHELM HIS, 1831–1904. AT SIXTY-FOUR YEARS.

out, and the palæontological records are searched, until the evidence from both sources establishes the fact that the organ or organism under study is but the summation of modifications and elaborations of a relatively simple primordial. This point settled, physiology is called upon to complete the story. Have the functions remained the same through the series? or have they undergone a series of modifications, differentiations, and improvements more or less parallel with the morphological series?"

Since physiology is an experimental science, all questions of this nature must be investigated with the help of experiments. Organisms undergoing development have been subjected to changed conditions, and their responses to various forms of stimuli have been noted. In the rise of experimental embryology we have one of the most promising of the recent departures from the older aspects of the subject. The results already attained in this attractive and suggestive field make too long a story to justify its telling in this volume. Roux, Herbst, Loeb, Morgan, E. B. Wilson, and many others have contributed to the growth of this new division of embryology. Good reasons have been adduced for believing that qualitative changes take place in the protoplasm as development proceeds. And a curb has been put upon that "great fault of embryology, the tendency to explain any and every operation of development as merely the result of inheritance." It has been demonstrated that surrounding conditions have much to do with individual development, and that the course of events may depend largely upon stimuli coming from without, and not exclusively on an inherited tendency.

Cell-Lineage.—Investigations on the structural side have reached a high grade of perfection in studies on cell-lineage. The theoretical conclusions in the germ-layer theory are based upon the assumption of identity in origin of the different layers. But the lack of agreement among observers, especially in reference to the origin of the mesoderm, made it necessary to study more closely the early developmental stages before the establishment of the germ-layers. It is a great triumph of exact observation that, although continually changing, the consecutive history of the individual cells has been followed from the beginning of segmentation to the time when the germ-layers are established. Some of the beautifully illustrated memoirs in this field are highly artistic.

Blochman (1882) was a pioneer in observations of this kind, and, following him, a number of American investigators have pursued studies on cell-lineage with great success. The researches of Whitman, Wilson, Conklin, Kofoid, Lillie, Mead, and Castle have given us the history of the origin of the germ-layers, cell by cell, in a variety of animal forms. These studies have shown that there is a lack of uniformity in the origin of at least the middle layer, and therefore there can be no strict homology of its derivatives. This makes it apparent that the earlier generalizations of the germ-layer theory were too sweeping, and, as a result, the theory is retained in a much modified form.

Theoretical Discussions.—Certain theoretical discussions, based on embryological studies, have been rife in recent years. And it is to be recognized without question that diseussions regarding heredity, regeneration, the nature of the developmental process, the question of inherited organization within the egg, of germinal continuity, etc., have done much to advance the subject of embryology.

Embryology is one of the three great departments of biology which, taken in combination, supply us with a knowledge of living forms along lines of structure, function, and development. The embryological method of study is of increasing importance to comparative anatomy and physiology. Formerly it was entirely structural, but it is now becoming also experimental, and it will therefore be of more service to physiology. While it has a strictly technical side, the science of embryology must always remain of interest to intelligent people as embracing one of the most wonderful processes in nature—the development of a complex organism from the single-celled condition, with a panoramic representation of all the intermediate stages.

CHAPTER XI

THE CELL THEORY—SCHLEIDEN, SCHWANN, SCHULTZE

THE recognition, in 1838, of the fact that all the various tissues of animals and plants are constructed on a similar plan was an important step in the rise of biology. It was progress along the line of microscopical observation. One can readily understand that the structural analysis of organisms could not be completed until their elementary parts had been discovered. When these units of structure were discovered they were called cells—from a misconception of their nature—and, although the misconception has long since been corrected, they still retain this historical but misleading name.

The doctrine that all tissues of animals and plants are composed of aggregations of these units, and the derivatives from the same, is known as the cell-theory. It is a generalization which unites all animals and plants on the broad plane of similitude of structure, and, when we consider it in the light of its consequences, it stands out as one of the great scientific achievements of the nineteenth century. There is little danger of overestimating the importance of this doctrine as tending to unify the knowledge of living organisms.

Vague Foreshadowings of the Cell-Theory.—In attempting to trace the growth of this idea, as based on actual observations, we first encounter vague foreshadowings of it in the seventeenth and the eighteenth centuries. The cells were seen and sketched by many early observers, but were not understood.

As long ago as 1665 Robert Hooke, the great English microscopist, observed the cellular construction of cork, and described it as made up of "little boxes or cells distinguished from one another." He made sketches of the appearance of this plant tissue; and, inasmuch as the drawings of Hooke are the earliest ones made of cells, they possess especial in-

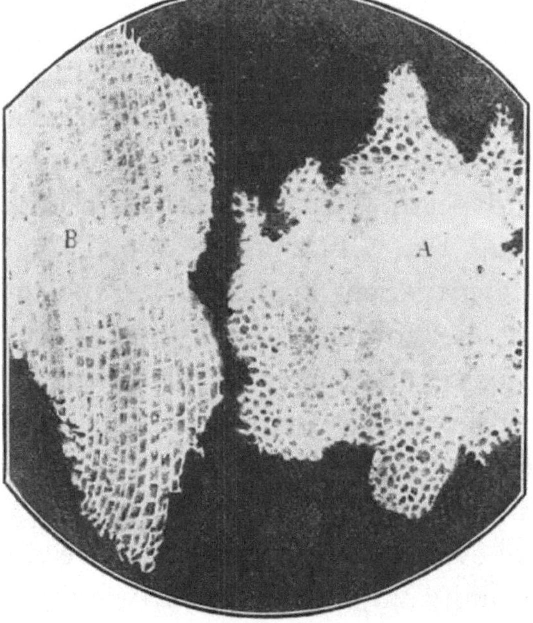

FIG. 72.—The Earliest Known Picture of Cells from Hooke's *Micrographia* (1665). From the edition of 1780.

terest and consequently are reproduced here. Fig. 72, taken from the *Micrographia*, shows this earliest drawing of Hooke. He made thin sections with a sharp penknife; "and upon examination they were found to be all cellular or porous in the manner of a honeycomb, but not so regular."

We must not completely overlook the fact that Aristotle (384–322 B.C.) and Galen (130–200 A.D.), those profound thinkers on anatomical structure, had reached the theoretical position "that animals and plants, complex as they may appear, are yet composed of comparatively few elementary parts, frequently repeated"; but we are not especially concerned with the remote history of the idea, so much as with the principal steps in its development after the beginning of microscopical observations.

Pictures of Cells in the Seventeenth Century.—The sketches illustrating the microscopic observations of Malpighi,

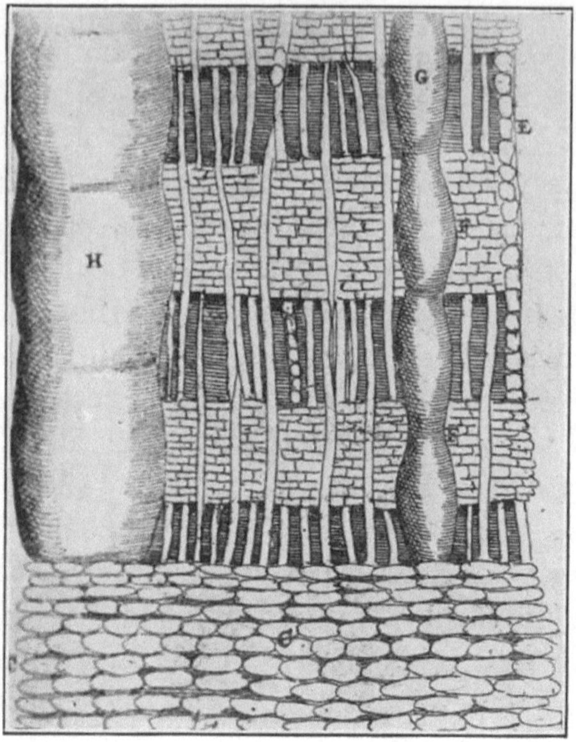

FIG. 73.—Sketch from Malpighi's Treatise on the Anatomy of Plants (1670).

Leeuwenhoek, and Grew show so many pictures of the cellular construction of plants that one who views them for the first time is struck with surprise, and might readily exclaim: "Here in the seventeenth century we have the foundation of the cell-theory." But these drawings were merely faithful representations of the appearance of the fabric of plants;

the cells were not thought of as uniform elements of organic architecture, and no theory resulted. It is true that Malpighi understood that the cells were separable "utricles," and that plant tissue was the result of their union, but this was only an initial step in the direction of the cell-theory, which, as we shall see later, was founded on the supposed identity in development of cells in animals and plants. Fig. 73 shows a sketch, made by Malpighi about 1670, illustrating the microscopic structure of a plant. This is similar to the many drawings of Grew and Leeuwenhoek illustrating the strueture of plant tissues.

Wolff.—Nearly a century after the work of Malpighi, we find Wolff, in 1759, proposing a theory regarding the organization of animals and plants based upon observations of their mode of development. He was one of the most acute scientific observers of the period, and it is to be noted that his conclusions regarding structure were all founded upon what he was able to see; while he gives some theoretical conclusions of a purely speculative nature, Wolff was careful to keep these separate from his observations. The purpose of his investigations was to show that there was no pre-formation in the embryo; but in getting at the basis of this question, he worked out the identity of structure of plants and animals as shown by their development. In his famous publication on the Theory of Development (*Theoria Generationis*) he used both plants and animals.

Huxley epitomizes Wolff's views on the development of elementary parts as follows: "Every organ, he says, is composed at first of a little mass of clear, viscous, nutritive fluid which possesses no organization of any kind, but is at most composed of globules. In this semifluid mass cavities (*Bläschen, Zellen*) are now developed; these, if they remain round or polygonal, become the subsequent cells; if they elongate, the vessels; and the process is identically the same, whether it is examined in the vegetating point of a plant, or in the young budding organs of an animal."

Wolff was contending against the doctrine of pre-formation in the embryo (see further under the chapter on Embryology), but on account of his acute analysis he should be regarded, perhaps, as the chief forerunner of the founders of the cell-theory. He contended for the same method of development that was afterward emphasized by Schleiden and Schwann. Through the opposition of the illustrious physiologist Haller his work remained unappreciated, and was finally forgotten, until it was revived again in 1812.

We can not show that Wolff's researches had any direct influence in leading Schleiden and Schwann to their announcement of the cell-theory. Nevertheless, it stands, intellectually, in the direct line of development of that idea, while the views of Haller upon the construction of organized beings are a side-issue. Haller declared that "the solid parts of animals and vegetables have this fabric in common, that their elements are either fibers or unorganized concrete." This formed the basis of the fiber-theory, which, on account of the great authority of Haller in physiology, occupied in the accumulating writings of anatomists a greater place than the views of Wolff.

Bichat, although he is recognized as the founder of histology, made no original observations on the microscopic units of the tissues. He described very minutely the membranes in the bodies of animals, but did not employ the microscope in his investigations.

Oken.—In the work of the dreamer Oken (1779–1851), the great representative of the German school of "*Naturphilosophie*," we find, about 1808, a very noteworthy statement to the effect that "animals and plants are throughout nothing else than manifoldly divided or repeated vesicles, as I shall prove anatomically at the proper time." This is apparently a concise statement of the cell-idea prior to Schleiden and Schwann; but we know that it was not founded on observation. Oken, as was his wont, gave rein to his imagination, and, on his part, the idea was entirely theoretical, and amounted to nothing more than a lucky guess.

Haller's fiber-theory gave place in the last part of the eighteenth century to the theory that animals and plants are composed of globules and formless material, and this globular theory was in force up to the time of the great generalization of Schleiden and Schwann. It was well expounded by Milne-Edwards in 1823, and now we can recognize that at least some of the globules which he described were the nucleated cells of later writers.

The Announcement of the Cell-Theory.—We are now approaching the time when the cell-theory was to be launched. During the first third of the nineteenth century there had accumulated a great mass of separate observations on the microscopic structure of both animals and plants. For several years botanists, in particular, had been observing and writing about cells, and interest in these structures was increasing. "We must clearly recognize the fact that for some time prior to 1838 the cell had come to be quite universally recognized as a constantly recurring element in vegetable and animal tissues, though little importance was attached to it as an element of organization, nor had its character been clearly determined" (Tyson).

Then, in 1838, came the "master-stroke in generalization" due to the combined labors of two friends, Schleiden and Schwann. But, although these two men are recognized as co-founders, thev do not share honors equally; the work of Schwann was much more comprehensive, and it was he who first used the term cell-theory, and entered upon the theoretical considerations which placed the theory before the scientific world.

Schleiden was educated as a lawyer, and began the practice of that profession, but his taste for natural science was so pronounced that when he was twenty-seven years old he deserted law, and went back to the university to study medicine. After graduating in medicine, he devoted himself mainly to botany. He saw clearly that the greatest thing needed for the advancement of scientific botany was a study of plant organization from the standpoint of development. Accordingly he entered upon this work, and, in 1837, arrived at a new view regarding the origin of plant cells. It must be confessed that this new view was founded on erroneous observations and conclusions, but it was revolutionary, and served to provoke discussion and to awaken observation. This was a characteristic feature of Schleiden's influence upon botany. His work acted as a ferment in bringing about new activity.

The discovery of the nucleus in plant cells by Robert Brown in 1831 was an important preliminary step to the work of Schleiden, since the latter seized upon the nucleus as the starting-point of new cells. He changed the name of the nucleus to cytoblast, and supposed that the new cell started as a small clear bubble on one side of the nucleus, and by continued expansion grew into the cell, the nucleus, or cytoblast, becoming encased in the cell-wall. All this was shown by Nägeli and other botanists to be wrong; yet, curiously enough, it was through the help of these false observations that Schwann arrived at his general conclusions.

Schleiden was acquainted with Schwann, and in October, 1838, while the two were dining together, he told Schwann about his observations and theories. He mentioned in particular the nucleus and its relationship to the other parts of the cell. Schwann was immediately struck with the similarity between the observations of Schleiden and certain of his own upon *animal* tissues. Together they went to his laboratory and examined the sections of the dorsal cord, the particular structure upon which Schwann had been working. Schleiden at once recognized the nuclei in this structure as being similar to those which he had observed in plants, and thus aided Schwann to come to the conclusion that the elements in animal tissues were practically identical with those in plant tissues.

Schwann.—The personalities of the co-founders of the cell-theory are interesting. Schwann was a man of gentle, pacific disposition, who avoided all controversies aroused by his many scientific discoveries. In his portrait (Fig. 74) we see a man whose striking qualities are good-will and benignity. His friend Henle gives this description of him: "He was a man of stature below the medium, with a beardless face, an almost infantile and always smiling expression, smooth, dark-brown hair, wearing a fur-trimmed dressing-gown, living in a poorly lighted room on the second floor of a restaurant which was not even of the second class. He would pass whole days there without going out, with a few rare books around him, and numerous glass vessels, retorts, vials, and tubes, simple apparatus which he made himself. Or I go in imagination to the dark and fusty halls of the Anatomical Institute where we used to work till nightfall by the side of our excellent chief, Johann Müller. We took our dinner in the evening, after the English fashion, so that we might enjoy more of the advantages of daylight."

Schwann drew part of his stimulus from his great master, Johannes Müller. He was associated with him as a student, first in the University of Würzburg, where Müller, with rare discernment for recognizing genius, selected Schwann for especial favors and for close personal friendship. The influence of his long association with Müller, the greatest of all trainers of anatomists and physiologists of the nineteenth century, must have been very uplifting. A few years later,

Schwann found himself at the University of Berlin, where Müller had been called, and he became an assistant in the master's laboratory. There he gained the powerful stimulus of constant association with a great personality.

FIG. 74.—THEODOR SCHWANN, 1810–1882.

In 1839, just after the publication of his work on the cell-theory, Schwann was called to a professorship in the University of Louvain, and after remaining there nine years, was transferred to the University of Liège. He was highly respected in the university, and led a useful life, although after going to Belgium he published only one work—that on the uses of the bile. He was recognized as an adept experi-

FIG. 75.—M. SCHLEIDEN, 1804–1881.

menter and demonstrator, and "clearness, order, and method" are designated as the characteristic qualities of his teaching.

His announcement of the cell-theory was his most important work. Apart from that his best-known contributions to science are: experiments upon spontaneous generation, his discovery of the "sheath of Schwann," in nerve fibers, and his theory of fermentation as produced by microbes.

Schleiden.—Schleiden (Fig. 75) was quite different in temperament from Schwann. He did not have the fine self-control of Schwann, but was quick to take up the gauntlet and enter upon controversies. In his caustic replies to his critics, he indulged in sharp personalities, and one is at times inclined to suspect that his early experience as a lawyer had something to do with his method of handling opposition. With all this he had correct ideas of the object of scientific study and of the methods to be used in its pursuit. He insisted upon observation and experiment, and upon the necessity of studying the development of plants in order to understand their anatomy and physiology. He speaks scornfullv of the botany of mere species-making as follows·

"Most people of the world, even the most enlightened, are still in the habit of regarding the botanist as a dealer in barbarous Latin names, as a man who gathers flowers, names them, dries them, and wraps them in paper, and all of whose wisdom consists in determining and classifying this hay which he has collected with such great pains."

Although he insisted on correct methods, his ardent nature led him to champion conclusions of his own before they were thoroughly tested. His great influence in the development of scientific botany lay in his earnestness, his application of new methods, and his fearlessness in drawing conclusions, which, although frequently wrong, formed the starting-point of new researches.

Let us now examine the original publications upon which the cell-theory was founded.

Schleiden's Contribution.—Schleiden's paper was particularly directed to the question, How does the cell originate?

and was published in Müller's *Archiv*, in 1838, under the German title of *Ueber Phytogenesis*. As stated above, the cell had been recognized for some years, but the question of its origin had not been investigated. Schleiden says: "I may omit all historical introduction, for, so far as I am acquainted, no direct observations exist at present upon the development of the cells of plants."

He then goes on to define his view of the nucleus (cytoblast) and of the development of the cell around it, saying: "As soon as the cytoblasts have attained their full size, a delicate transparent vesicle arises upon their surface. This is the young cell." As to the position of the nucleus in the fully developed cell, he is very explicit: "It is evident," he says, "from the foregoing that the cytoblast can never lie free in the interior of the cell, but is always enclosed in the cell-wall," etc.

Schleiden fastened these errors upon the cell-theory, since Schwann relied upon his observations. On another point of prime importance Schleiden was wrong: he regarded all new cell-formation as the formation of "cells within cells," as distinguished from cell-division, as we now know it to take place.

Schleiden made no attempt to elaborate his views into a comprehensive cell-theory, and therefore his connection as a co-founder of this great generalization is chiefly in paving the way and giving the suggestion to Schwann, which enabled the latter to establish the theory. Schleiden's paper occupies some thirty-two pages, and is illustrated by two plates. He was thirty-four years old when this paper was published, and directly afterward was called to the post of adjunct professor of botany in the University of Jena, a position which with promotion to the full professorship he occupied for twenty-three years.

Schwann's Treatise.—In 1838, Schwann also announced his cell-theory in a concise form in a German scientific periodical, and, later, to the Paris Academy of Sciences; but it was not till 1839 that the fully illustrated account was published. This treatise with the cumbersome title, "Microscopical Researches into the Accordance in the Structure and Growth of Animals and Plants" (*Mikroscopische Untersuchungen über die Uebereinstimmung in der Structur und dem Wachsthum der Thiere und Pflanzen*) takes rank as one of the great classics in biology. It fills 215 octavo pages, and is illustrated with four plates.

"The purpose of his researches was to prove the identity of structure, as shown by their development, between animals and plants." This is done by direct comparisons of the elementary parts in the two kingdoms of organic nature.

His writing in the "Microscopical Researches" is clear and philosophical, and is divided into three sections, in the first two of which he confines himself strictly to descriptions of observations, and in the third part of which he enters upon a philosophical discussion of the significance of the observations. He comes to the conclusion that "the elementary parts of all tissues are formed of cells in an analogous, though very diversified manner, so that it may be asserted that there is one universal principle of development for the elementary parts of organisms, however different, and that this principle is the formation of cells."

It was in this treatise also that he made use of the term cell-theory, as follows: "The development of the proposition that there exists one general principle for the formation of all organic productions, and that this principle is the formation of cells, as well as the conclusions which may be drawn from this proposition, may be comprised under the term *cell-theory*, using it in its more extended signification, while, in a more limited sense, by the theory of cells we understand whatever may be inferred from this proposition with respect to the powers from which these phenomena result."

One comes from the reading of these two contributions to science with the feeling that it is really Schwann's cell-theory, and that Schleiden helped by lighting the way that his fellow-worker so successfully trod.

Modification of the Cell-Theory.—The form in which the cell-theory was given to the world by Schleiden and Schwann was very imperfect, and, as already pointed out, it contained fundamental errors. The founders of the theory attached too much importance to the cell-wall, and they described the cell as a hollow cavity bounded by walls that were formed around a nucleus. They were wrong as to the mode of the development of the cell, and as to its nature. Nevertheless, the great truth that all parts of animals and plants are built of similar units or structures was well substantiated. This remained a permanent part of the theory, but all ideas regarding the nature of the units were profoundly altered.

In order to perceive the line along which the chief modifications were made we must take account of another scientific advance of about the same period. This was the discovery of protoplasm, an achievement which takes rank with the advances of greatest importance in biology, and has proved to be one of the great events of the nineteenth century.

The Discovery of Protoplasm and its Effect on the Cell Theory.—In 1835, before the announcement of the cell-theory, living matter had been observed by Dujardin. In lower animal forms he noticed a semifluid, jelly-like substance, which he designated sarcode, and which he described as being endowed with all the qualities of life. The same semifluid substance had previously caught the attention of some observers, but no one had as yet announced it as the actual living part of organisms. Schleiden had seen it and called it gum. Dujardin was far from appreciating the full importance of his discovery, and for a long time his description of sarcode remained separate; but in 1846 Hugo von

Mohl, a botanist, observed a similar jelly-like substance in plants, which he called plant *schleim*, and to which he attached the name protoplasma.

The scientific world was now in the position of recognizing living substance, which had been announced as sarcode in lower animals, and as protoplasm in plants; but there was as yet no clear indication that these two substances were practically identical. Gradually there came stealing into the minds of observers the suspicion that the sarcode of the zoölogists and the protoplasm of the botanists were one and the same thing. This proposition was definitely maintained by Cohn in 1850, though with him it was mainly theoretical, since his observations were not sufficiently extensive and accurate to support such a conclusion.

Eleven years later, however, as the result of extended researches, Max Schultze promulgated, in 1861, the protoplasm doctrine, to the effect that the units of organization consist of little masses of protoplasm surrounding a nucleus, and that this protoplasm, or living substance, is practically identical in both plants and animals.

The effect of this conclusion upon the cell-theory was revolutionary. During the time protoplasm was being observed the cell had likewise come under close scrutiny, and naturalists had now an extensive collection of facts upon which to found a theory. It has been shown that many animal cells have no cell-wall, and the final conclusion was inevitable that the essential part of a cell is the semifluid living substance that resides within the cavity when a cell-wall is present. Moreover, when the cell-wall is absent, the protoplasm is the "cell." The position of the nucleus was also determined to be within the living substance, and not, as Schleiden had maintained, within the cell-wall. The definition of Max Schultze, that a cell is a globule of protoplasm surrounding a nucleus, marks a new era in the cell-theory, in which the original generalization became consolidated with the protoplasm doctrine.

Further Modifications of the Cell-Theory.—The reformed cell-theory was, however, destined to undergo further modification, and to become greatly extended in its application. At first the cell was regarded merely as an element of structure; then, as a supplement to this restricted view, came the recognition that it is also a unit of physiology, *viz.*, that all physiological activities take place within the cell. Matters did not come to a rest, however, with the recognition of these two fundamental aspects of the cell. The importance of the cell in development also took firmer hold upon the minds of anatomists after it was made clear that both the egg and its fertilizing agents are modified cells of the parent's body. It was necessary to comprehend this fact in order to get a clear idea of the origin of cells within the body of a multicellular organism, and of the relation between the primordial element and the fully developed tissues. Finally, when observers found within the nucleus the bearers of hereditary qualities, they began to realize that a careful study of the behavior of the cell elements during development is necessary for the investigation of hereditary transmissions.

A statement of the cell-theory at the present time, then, must include these four conceptions: the cell as a unit of structure, the cell as a unit of physiological activity, the cell as embracing all hereditary qualities within its substance, and the cell in the historical development of the organism.

Some of these relations may now be more fully illustrated.

Origin of Tissues.—The egg in which all organisms above the very lowest begin, is a single cell having, under the microscope, the appearance shown in Fig. 76. After fertilization, this divides repeatedly, and many cohering cells result. The cells are at first similar, but as they increase in number, and as development proceeds, they grow different, and certain groups are set apart to perform particular duties. The division of physiological labor which arises at this time marks the beginning of separate tissues. It has been demonstrated over and over that all tissues are composed of cells and cell-products, though in some instances they are much modified. The living cells can be seen even in bone and cartilage, in

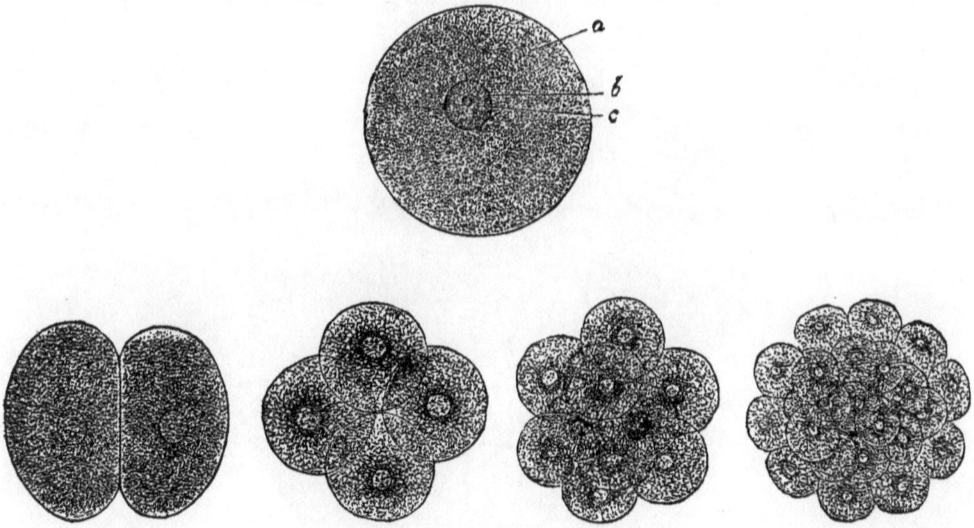

FIG. 76.—The Egg and Early Stages in its Development. (After Gegenbaur.)

which they are separated by a lifeless matrix, the latter being the product of cellular activity.

Fig. 77 shows a stage in the development of one of the mollusks just as the differentiation of cells has commenced.

The Nucleus.—To the earlier observers the protoplasm appeared to be a structureless, jelly-like mass containing granules and vacuoles; but closer acquaintance with it has shown that it is in reality very complex in structure as well as in chemical composition. It is by no means homogeneous; adjacent parts are different in properties and aptitudes. The nucleus, which is more readily seen than other cell elements, was shown to be of great importance in cell-life—to be a structure which takes the lead in cell division, and in general dominates the rest of the protoplasm.

Chromosomes.—After dyes came into use for staining the protoplasm (1868), it became evident that certain parts of it stain deeply, while other parts stain faintly or not at all. This led to the recognition of protoplasm as made up of a densely staining portion called *chromatin*, and a faintly staining por

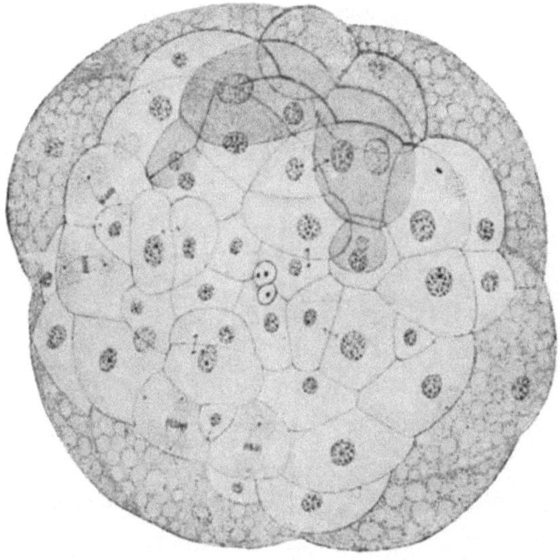

FIG. 77.—An Early Stage in the Development of the Egg of a Rock-Limpet (After Conklin.)

tion designated *achromatin*. This means of making different parts of protoplasm visible under the microscope led to important results, as when, in 1883, it was discovered that the nucleus contains a definite number of small (usually rod-shaped) bodies, which become evident during nuclear division, and play a wonderful part in that process. These bodies take the stain more deeply than other components of the nucleus, and are designated *chromosomes*.

Attention having been directed to these little bodies, continued observations showed that, although they vary in number—commonly from two to twenty-four—in different parts of animals and plants, they are, nevertheless, of the same number in all the cells of any particular plant or ani-

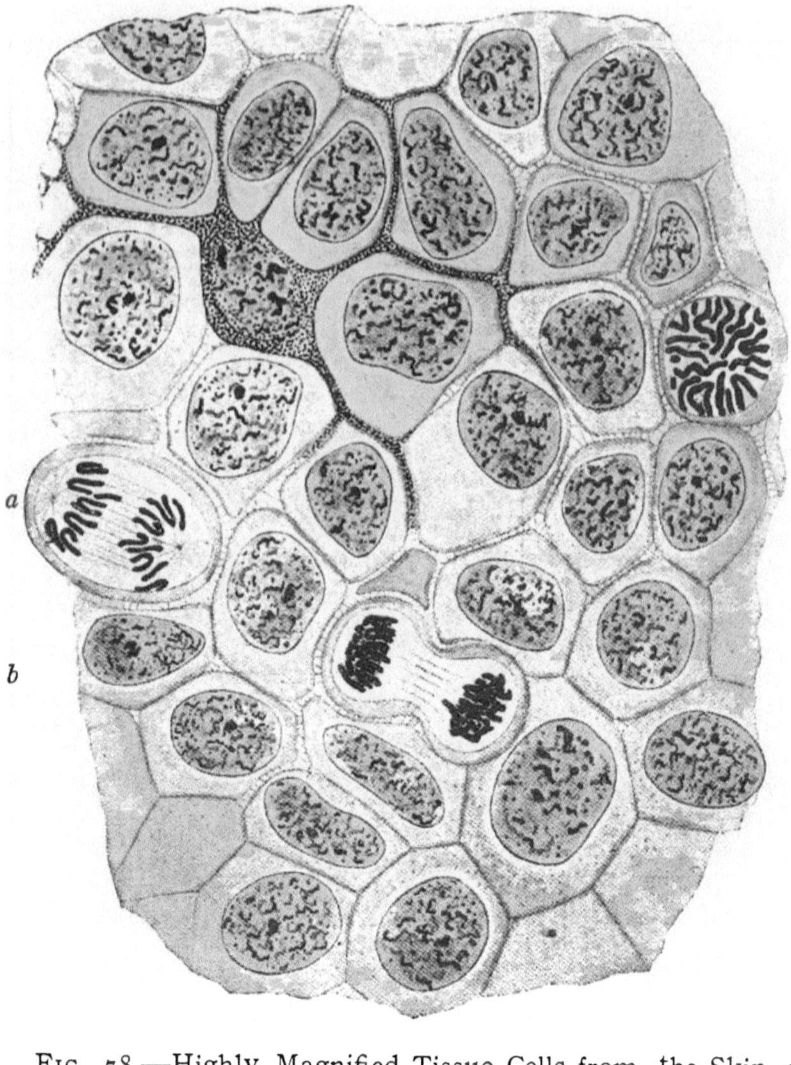

FIG. 78.—Highly Magnified Tissue Cells from the Skin of a Salamander in an Active State of Growth. Dividing cells with chromosomes are shown at *a*, *b*, and *c*,. (After Wilson.)

mal. As a conclusion to this kind of observation, it needs to be said that the chromosomes are regarded as the actual bearers of hereditary qualities. The chromosomes do not show in resting-stages of the nucleus; their substance is present, but is not aggregated into the form of chromosomes.

Fig. 78 shows tissue cells, some of which are in the divid ing and others in the resting-stage. The nuclei in process of

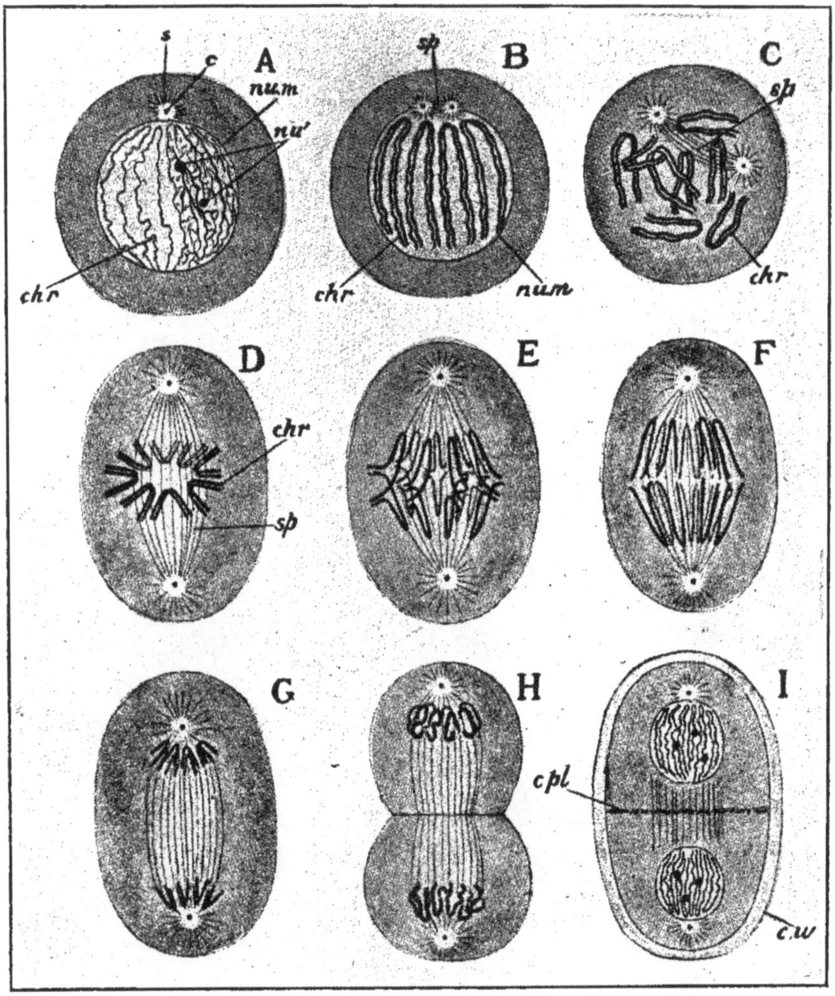

FIG. 79.—Diagram of the Chief Steps in Cell-division. (After Parker as altered from Fleming.)

division exhibit the rod-like chromosomes, as shown at *a*, *b*, and *c*.

Centrosome.—The discovery (1876) of a minute spot of deeply staining protoplasm, usually just outside the nuclear membrane, is another illustration of the complex structure of the cell. Although the centrosome, as this spot is called, has been heralded as a dynamic agent, there is not complete agreement as to its purpose, but its presence makes it necessary to include it in the definition of a cell.

The Cell in Heredity.—The problems of inheritance, in so far as they can be elucidated by structural studies, have come to be recognized as problems of cellular life. But we cannot understand what is implied by this conclusion without referring to the behavior of the chromosomes during cell-division. This is a very complex process, and varies somewhat in different tissues. We can, however, with the help of Fig. 79, describe what takes place in a typical case. The nucleus does not divide directly, but the chromosomes congregate around the equator of a spindle (*D*) formed from the achromatin; they then undergo division lengthwise, and migrate to the poles (*E*, *F*, *G*), after which a partition wall is formed dividing the cell. This manner of division of the chromosomes secures an equable partition of the protoplasm. In the case of fertilized eggs, one-half of the chromosomes are derived from the sperm and one-half from the egg. Each cell thus contains hereditary substance derived from both maternal and paternal nuclei. This is briefly the basis for regarding inheritance as a phenomenon of cell-life.

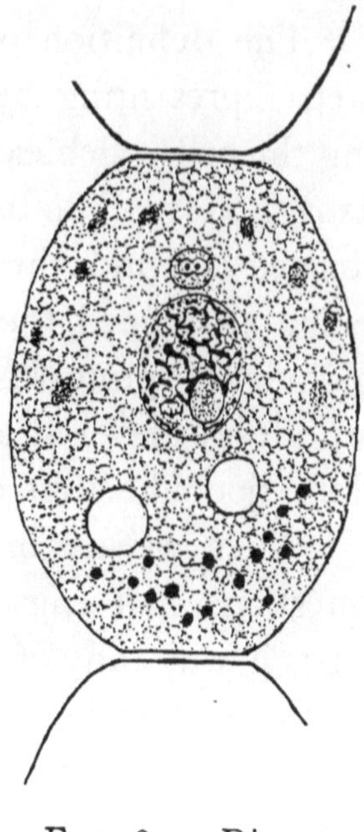

FIG. 80.—Diagram of a Cell. (Modified after Wilson.)

A diagram of the cell as now understood (Fig. 80) will be helpful in showing how much the conception of the cell has changed since the time of Schleiden and Schwann.

Definition.—The definition of Verworn, made in 1895, may be combined with this diagram: A cell is "a body consisting essentially of protoplasm in its general form, including the unmodified cytoplasm, and the specialized nucleus and centrosome; while as unessential accompaniments may be enumerated: (1) the cell membrane, (2) starch grains, (3) pigment granules, (4) oil globules, and (5) chlorophyll granules." No definition can include all variations, but the one quoted is excellent in directing attention to the essentials—to protoplasm in its general form, and the modified protoplasmic parts as distinguished from the unessential aecompaniments, as cell membrane and cell contents.

The definition of Verworn was reached by a series of steps representing the historical advance of knowledge regarding the cell. Schleiden and Schwann looked upon the cell as a hollow chamber having a cell-wall which had been formed around the nucleus; it was a great step when Schultze defined the cell in terms of living substance as "a globule of protoplasm surrounding a nucleus," and it is a still deeper level of analysis which gives us a discriminating definition like that of Verworn.

When we are brought to realize that, in large part, the questions that engage the mind of the biologist have their basis in the study of cells, we are ready to appreciate the force of the statement that the establishment of the cell-theory was one of the great events of the nineteenth century, and, further, that it stands second to no theory, with the single exception of that of organic evolution, in advancing biological science.

CHAPTER XII

PROTOPLASM, THE PHYSICAL BASIS OF LIFE

THE recognition of the rôle that protoplasm plays in the living world was so far-reaching in its results that we take up for separate consideration the history of its discovery. Although it is not yet fifty years since Max Schultze established the protoplasm doctrine, it has already had the greatest influence upon the progress of biology. To the consideration of protoplasm in the previous chapter should be added an account of the conditions of its discovery, and of the personality and views of the men whose privilege it was to bring the protoplasm idea to its logical conclusion. Before doing so, however, we shall look at the nature of protoplasm itself.

Protoplasm.—This substance, which is the seat of all vital activity, was designated by Huxley "the physical basis of life," a graphic expression which brings before the mind the central fact that life is manifested in a material substratum by which it is conditioned. All that biologists have been able to discover regarding life has been derived from the observation of that material substratum. It is not difficult, with the help of a microscope, to get a view of protoplasmic activity, and that which was so laboriously made known about 1860 is now shown annually to students beginning biology

Inasmuch as all living organisms contain protoplasm, one has a wide range of choice in selecting the plant or the animal upon which to make observations.

We may, for illustration, take one of the simplest of animal organisms, the amœba, and place it under the high powers of the microscope. This little animal consists almost entirely of a lump of living jelly. Within the living substance of which its body is composed all the vital activities characteristic of higher animals are going on, but they are manifested in simpler form. These manifestations differ only in degree of development, not in kind, from those we see in bodies of higher organisms.

We can watch the movements in this amœba, determine at first hand its inherent qualities, and then draw up a sort of catalogue of its vital properties. We notice an almost continual flux of the viscid substance, by means of which it is able to alter its form and to change its position. This quality is called that of contractility. In its essential nature it is like the protoplasmic movement that takes place in a contracting muscle. We find also that the substance of the amœba responds to stimulations—such as touching it with a bristle, or heating it, or sending through it a light electric shock. This response is quite independent of the contractility, and by physiologists is designated the property of being irritable.

By further observations one may determine that the substance of the amœba is receptive and assimilative, that it is respiratory, taking in oxygen and giving off carbonic dioxide, and that it is also secretory. If the amœba be watched long enough, it may be seen to undergo division, thus producing another individual of its kind. We say, therefore, that it exhibits the power of reproduction. All these properties manifested in close association in the amœba are exhibited in the bodies of higher organisms in a greater degree of perfection, and also in separation, particular organs often being set apart for the performance of one of these particular functions. We should, however, bear in mind that in the simple protoplasm of the amœba is found the germ of all the activities of the higher animals.

It will be convenient now to turn our attention to the microscopic examination of a plant that is sufficiently transparent to enable us to look within its living parts and observe the behavior of protoplasm. The first thing that strikes one is the continual activity of the living substance within the boundaries of a particular cell. This movement sometimes

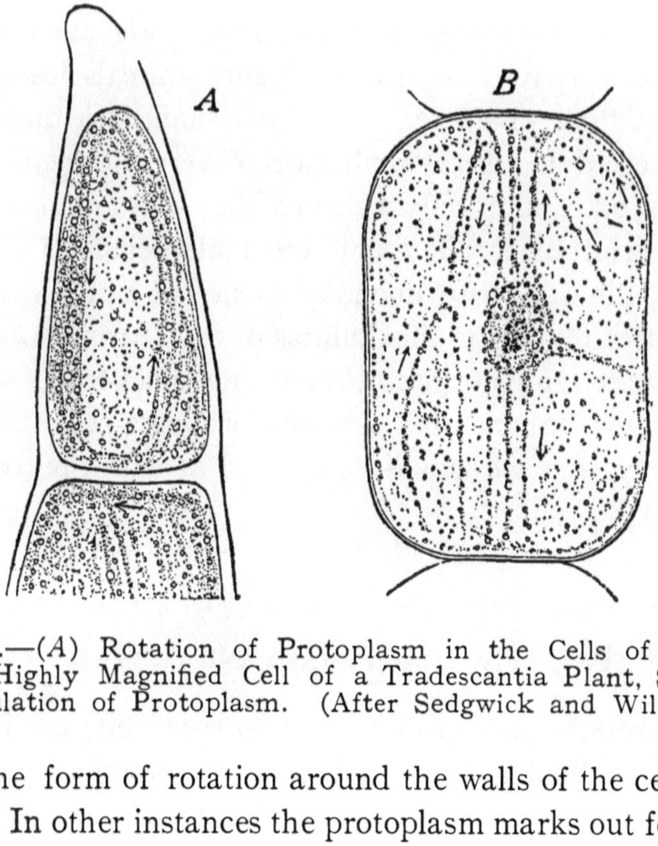

FIG. 81.—(*A*) Rotation of Protoplasm in the Cells of Nitella. (*B*) Highly Magnified Cell of a Tradescantia Plant, Showing Circulation of Protoplasm. (After Sedgwick and Wilson.)

takes the form of rotation around the walls of the cell (Fig. 81 *A*). In other instances the protoplasm marks out for itself new paths, giving a more complicated motion, called circulation (Fig. 81 *B*). These movements are the result of chemical changes taking place within the protoplasm, and they are usually to be observed in any plant or animal organism.

Under the most favorable conditions these movements, as seen under the microscope, make a perfect torrent of unceasing activity, and introduce us to one of the wonderful sights of which students of biology have so many. Huxley (with slight verbal alterations) says: "The spectacle afforded by the wonderful energies imprisoned within the compass of the microscopic cell of a plant, which we commonly regard as a merely passive organism, is not easily forgotten by one who has watched its movement hour by hour without pause or sign of weakening. The possible complexity of many other organisms seemingly as simple as the protoplasm of the plant just mentioned dawns upon one, and the comparison of such activity to that of higher animals loses much of its startling character. Currents similar to these have been observed in a great multitude of very different plants, and it is quite uniformly believed that they occur in more or less perfection in all young vegetable cells. If such be the case, the wonderful noonday silence of a tropical forest is due, after all, only to the dullness of our hearing, and could our ears catch the murmur of these tiny maelstroms as they whirl in the innumerable myriads of living cells that constitute each tree, we should be stunned as with the roar of a great city."

The Essential Steps in Recognizing the Likeness of Protoplasm in Plants and Animals

Dujardin.—This substance, of so much interest and importance to biologists, was first clearly described and distinguished from other viscid substance, as albumen, by Félix Dujardin in 1835. Both the substance and the movements therein had been seen and recorded by others: by Rösel von Rosenhof in 1755 in the proteus animalcule; again in 1772 by Corti in chara; by Mayen in 1827 in Vallisnieria; and in 1831 by Robert Brown in Tradescantia. One of these records was for the animal kingdom, and three were for plants. The observations of Dujardin, however, were on a different plane from those of the earlier naturalists, and he is usually credited with being the discoverer of protoplasm. His researches, moreover, were closely connected with the development of the ideas regarding the rôle played in nature by this living substance.

Dujardin was a quiet modest man, whose attainments and service to the progress of biology have usually been underrated. He was born in 1801 at Tours, and died in 1860 at Rennes. Being descended from a race of watchmakers, he received in his youth a training in that craft which cultivated his natural manual dexterity, and, later, this assisted him in his manipulations of the microscope. He had a fondness for sketching, and produced some miniatures and other works of art that showed great merit. His use of colors was very effective, and in 1818 he went to Paris for the purpose of perfecting himself in painting, and with the intention of becoming an artist. The small financial returns, however, "led him to accept work as an engineer directing the construction of hydraulic work in Sédan." He had already shown a love for natural science, and this led him from engineering into work as a librarian and then as a teacher. He made field observations in geology and botany, and commenced publication in those departments of science.

About 1834 he began to devote his chief efforts to microscopic work, toward which he had a strong inclination, and from that time on he became a zoölogist, with a steadily growing recognition for high-class observation. Besides his technical scientific papers, he wrote in a popular vein to increase his income. Among his writings of this type may be mentioned as occupying high rank his charmingly written "Rambles of a Naturalist" (*Promenades d'un Naturaliste*, 1838).

By 1840 he had established such a good record as a scientific investigator that he was called to the newly founded University of Rennes as dean of the faculty. He found himself in an atmosphere of jealous criticism, largely on account of his being elevated to the station of dean, and after two years of discomfort he resigned the deanship, but retained his position as a professor in the university. He secured a residence in a retired spot near a church, and lived there simply. In his leisure moments he talked frequently with the priests, and became a devout Catholic.

His contributions to science cover a wide range of subjects. In his microscopic work he discovered the rhizopods in 1834, and the study of their structure gave him the key to that of the other protozoa. In 1835 he visited the Mediterranean, where he studied the oceanic foraminifera, and demonstrated that they should be grouped with the protozoa, and not, as had been maintained up to that time, with the mollusca. It was during the prosecution of these researches that he made the observations upon sarcode that are of particular interest to us.

His natural history of the infusoria (1841) makes a volume of 700 pages, full of original observations and sketches. He also invented a means of illumination for the microscope, and wrote a manual of microscopic observation. Among the ninety-six publications of Dujardin listed by Professor Joubin there are seven general works, twenty relating to the protozoa, twenty-four to geology, three to botany, four to physics, twenty-five to arthropods, eight to worms, etc., etc. But as Joubin says: "The great modesty of Dujardin allowed him to see published by others, without credit to himself, numerous facts and observations which he had established." This failure to assert his claims accounts in part for the inadequate recognition that his work has received.

No portrait of Dujardin was obtainable prior to 1898. Somewhat earlier Professor Joubin, who succeeded other occupants of the chair which Dujardin held in the University of Rennes, found in the possession of his descendants a portrait, which he was permitted to copy. The earliest reproduction of this picture to reach this country came to the

FIG. 82.—FÉLIX DUJARDIN, 1801–1860.

writer through the courtesy of Professor Joubin, and a copy of it is represented in Fig. 82. His picture bespeaks his personality. The quiet refinement and sincerity of his face are evident. Professor Joubin published, in 1901 (*Archives de Parasitologie*), a biographical sketch of Dujardin, with several illustrations, including this portrait and another one which is very interesting, showing him in academic costume. Thanks to the spread of information of the kind contained in that article, Dujardin is coming into wider recognition, and will occupy the historical position to which his researches entitle him.

It was while studying the protozoa that he began to take particular notice of the substance of which their bodies are composed; and in 1835 he described it as a living jelly endowed with all the qualities of life. He had seen the same jelly-like substance exuding from the injured parts of worms, and recognized it as the same material that makes the body of protozoa. He observed it very carefully in the ciliated infusoria—in Paramœcium, in Vorticella, and other forms, but he was not satisfied with mere microscopic observation of its structure. He tested its solubility, he subjected it to the action of alcohol, nitric acid, potash, and other chemical substances, and thereby distinguished it from albumen, mucus, gelatin, etc.

Inasmuch as this substance manifestly was soft, Dujardin proposed for it the name of sarcode, from the Greek, meaning *soft*. Thus we see that the substance protoplasm was for the first time brought very definitely to the attention of naturalists through the study of animal forms. For some time it occupied a position of isolation, but ultimately became recognized as being identical with a similar substance that occurs in plants. At the time of Dujardin's discovery, sarcode was supposed to be peculiar to lower animals; it was not known that the same substance made the living part of all animals, and it was owing mainly to this circumstance that the full recognition of its importance in nature was delayed.

The fact remains that the first careful studies upon sarcode were due to Dujardin, and, therefore, we must include him among the founders of modern biology.

Purkinje.—The observations of the Bohemian investigator Purkinje (1787–1869) form a link in the chain of events leading up to the recognition of protoplasm. Although Purkinje is especially remembered for other scientific contri-

FIG. 83.—PURKINJE, 1787–1869.

butions, he was the first to make use of the name protoplasm for living matter, by applying it to the formative substance within the eggs of animals and within the cells of the embryo. His portrait is not frequently seen, and, therefore, is included here (Fig. 83), to give a more complete series of pictures of the men who were directly connected with the development of the protoplasm idea. Purkinje was successively a professor in the universities of Breslau and Prague. His anatomical laboratory at Breslau is notable as being one of the earliest (1825) open to students. He went to Prague in 1850 as professor of physiology.

Von Mohl.—In 1846, eleven years after the discovery of Dujardin, the eminent botanist Hugo von Mohl (1805–1872) designated a particular part of the living contents of the vegetable cell by the term protoplasma. The viscid, jelly-like

FIG. 84.—CARL NÄGELI, 1817–1891.

substance in plants had in the mean time come to be known under the expressive term of plant "*schleim.*" He distinguished the firmer mucilaginous and granular constituent, found just under the cell membrane, from the watery cell-sap that occupies the interior of the cell. It was to the former part that he gave the name protoplasma. Previous to this, the botanist Nägeli had studied this living substance, and perceived that it was nitrogenous matter. This was a distinct step in advance of the vague and indefinite idea of Schleiden, who had in reality noticed protoplasm in 1838, but thought of it merely as gum. The highly accomplished investigator Nägeli (Fig. 84) made a great place for himself

FIG. 85.—HUGO VON MOHL, 1805–1872.

in botanical investigation, and his name is connected with several fundamental ideas of biology. To Von Mohl, however, belongs the credit of having brought the word proto plasm into general use. He stands in the direct line of development, while Purkinje, who first employed the word protoplasm, stands somewhat aside, but his name, nevertheless, should be connected with the establishment of the protoplasm doctrine.

Von Mohl (Fig. 85) was an important man in botany. Early in life he showed a great love for natural science, and as in his day medical instruction afforded the best opportunities for a man with scientific tastes, he entered upon that course of study in Tübingen at the age of eighteen. He took his degree of doctor of medicine in 1823, and spent several years in Munich. He became professor of physiology in Bern in 1832, and three years later was transferred to Tübingen as professor of botany. Here he remained to the end of his life, refusing invitations to institutions elsewhere. He never married, and, without the cares and joys of a family, led a solitary and uneventful life, devoted to botanical investigation.

Cohn.—After Von Mohl's studies on "plant schleim" there was a general movement toward the conclusion that the sarcode of the zoölogists and the protoplasm of the botanists were one and the same substance. This notion was in the minds of more than one worker, but it is perhaps to Ferdinand Cohn (1828–1898) that the credit should be given for bringing the question to a head. After a study of the remarkable movements of the active spores of one of the simplest plants (protococcus), he said that vegetable protoplasm and animal sarcode, "if not identical, must be, at any rate, in the highest degree analogous substances" (Geddes).

Cohn (Fig. 86) was for nearly forty years professor of botany in the University of Breslau, and during his long life as an investigator greatly advanced the knowledge of bacteria. His statement referred to above was made when he was twenty-two years of age, and ran too far ahead of the evidence then accumulated; it merely anticipated the coming period of the acceptance of the conclusion in its full significance.

De Bary.—We find, then, in the middle years of the nineteenth century the idea launched that sarcode and protoplasm are identical, but it was not yet definitely established

FIG. 86.—FERDINAND COHN, 1828–1898.

that the sarcode of lower animals is the same as the living substance of the higher ones, and there was, therefore, lacking an essential factor to the conclusion that there is only one general form of living matter in all organisms. It took another ten years of investigation to reach this end.

The most important contributions from the botanical side during this period were the splendid researches of De Bary (Fig. 87) on the myxomycetes, published in 1859. Here the resemblance between sarcode and protoplasm was brought out with great clearness. The myxomycetes are, in one condition, masses of vegetable protoplasm, the movements and other characteristics of which were shown to resemble strongly those of the protozoa. De Bary's great fame as a botanist has made his name widely known.

In 1858 Virchow also, by his extensive studies in the pathology of living cells, added one more link to the chain

FIG. 87.—HEINRICH A. DE BARY, 1831–1888.

that was soon to be recognized as encircling the new domain of modern biology.

Schultze.—As the culmination of a long period of work, Max Schultze, in 1861, placed the conception of the identity between animal sarcode and vegetable protoplasm upon an unassailable basis, and therefore he has received the title of

FIG. 88.—MAX SCHULTZE, 1825–1874.

"the father of modern biology." He showed that sarcode, which was supposed to be confined to the lower invertebrates, is also present in the tissues of higher animals, and there exhibits the same properties. The qualities of contractility and irritability were especially indicated. It was on physiological likeness, rather than on structural grounds, that he formed his sweeping conclusions. He showed also that sarcode agreed in physiological properties with protoplasm in plants, and that the two living substances were practically identical. His paper of 1861 considers the living substance in muscles (*Ueber Muskelkörperchen und das was man eine Zelle zu nennen habe*), but in this he had been partly anticipated by Ecker who, in 1849, compared the "formed contractile substance" of muscles with the "unformed contractile substance" of the lower types of animal life (Geddes).

The clear-cut, intellectual face of Schultze (Fig. 88) is that of an admirable man with a combination of the artistic and the scientific temperaments. He was greatly interested in music from his youth up, and by the side of his microscope was his well-beloved violin. He was some time professor in the University of Halle, and in 1859 went to Bonn as professor of anatomy and director of the Anatomical Institute. His service to histology has already been spoken of (Chapter VIII).

This astute observer will have an enduring fame in biological science, not only for the part he played in the development of the protoplasm idea, but also on account of other extensive labors. In 1866 he founded the leading periodical in microscopic anatomy, the *Archiv für Mikroscopische Anatomie.* This periodical was continued after the untimely death of Schultze in 1874, and to-day is one of the leading biological periodicals.

It is easy, looking backward, to observe that the period between 1840 and 1860 was a very important one for modern biology. Many new ideas were coming into existence, but through this period we can trace distinctly, step by step, the gradual approach to the idea that protoplasm, the living substance of organism, is practically the same in plants and in animals. Let us picture to ourselves the consequences of the acceptance of this idea. Now for the first time physiologists began to have their attention directed to the actually living substance; now for the first time they saw clearly that all future progress was to be made by studying this living substance—the seat of vital activity. This was the beginning of modern biology.

Protoplasm is the particular object of study for the biologist. To observe its properties, to determine how it behaves under different conditions, how it responds to stimuli and natural agencies, to discover the relation of the internal changes to the outside agencies: these, which constitute the fundamental ideas of biology, were for the first time brought directly to the attention of the naturalist, about the year 1860—that epoch-making time when appeared Darwin's *Origin of Species* and Spencer's *First Principles.*

CHAPTER XIII

THE WORK OF PASTEUR, KOCH, AND OTHERS

THE knowledge of bacteria, those minutest forms of life, has exerted a profound influence upon the development of general biology. There are many questions relating to bacteria that are strictly medical, but other phases of their life and activities are broadly biological, and some of those broader aspects will next be brought under consideration.

The bacteria were first described by Leeuwenhoek in 1687, twelve years after his discovery of the microscopic animalcula now called protozoa. They are so infinitesimal in size that under his microscope they appeared as mere specks, and, naturally, observation of these minute organisms was suspended until nearly the middle of the nineteenth century, after the improvement of microscope lenses. It is characteristic of the little knowledge of bacteria in Linnæus's period that he grouped them into an order, with other microscopic forms, under the name *chaos*.

At first sight, the bacteria appear too minute to figure largely in human affairs, but a great department of natural science—bacteriology—has been opened by the study of their activities, and it must be admitted that the development of the science of bacteriology has been of great practical importance. The knowledge derived from experimental studies of the bacteria has been the chief source of light in an obscure domain which profoundly affects the well-being of mankind. To the advance of such knowledge we owe the germ-theory of disease and the ability of medical men to cope with contagious diseases. The three greatest names connected with the rise of bacteriology are those of Pasteur, Koch, and Lister, the results of whose labors will be considered later.

Among the general topics which have been clustered around the study of bacteria we take up, first, the question of the spontaneous origin of life.

The Spontaneous Origin of Life

It will be readily understood that the question of the spontaneous generation of life is a fundamental one for the biologist.. Does life always arise from previously existing life, or under certain conditions is it developed spontaneously? Is there, in the inorganic world, a happy concourse of atoms that become chained together through the action of the sun's rays and other natural forces, so that a molecule of living matter is constructed in nature's laboratory without contact or close association with living substance? This is a question of *biogenesis*—life from previous life—or of *abiogenesis*—life without preëxisting life or from inorganic matter alone.

It is a question with a long history. Its earliest phases do not involve any consideration of microscopic forms, since they were unknown, but its middle and its modern aspect are concerned especially with bacteria and other microscopic organisms. The historical development of the problem may be conveniently considered under three divisions: I. The period from Aristotle, 325 B.C., to the experiments of Redi, in 1668; II. From the experiments of Redi to those of Schulze and Schwann in 1836 and 1837; III. The modern phase, extending from Pouchet's observations in 1859 to the present.

I. From Aristotle to Redi.—During the first period, the notion of spontaneous generation was universally accepted, and the whole question of spontaneous origin of life was in a crude and grotesque condition. It was thought that frogs and toads and other animals arose from the mud of ponds and streams through the vivifying action of the sun's rays. Rats were supposed to come from the river Nile, the dew was supposed to give origin to insects, etc.

The scientific writers of this period had little openness of mind, and they indulged in scornful and sarcastic comments at the expense of those who doubted the occurrence of spontaneous generation. In the seventeenth century Alexander Ross, commenting on Sir Thomas Brown's doubt as to whether mice may be bred by putrefaction, flays his antagonist in the following words: "So may we doubt whether in cheese and timber worms are generated, or if beetles and wasps in cow-dung, or if butterflies, locusts, shell-fish, snails, eels, and such life be procreated of putrefied matter, which is to receive the form of that creature to which it is by formative power disposed. To question this is to question reason, sense, and experience. If he doubts this, let him go to Egypt, and there he will find the fields swarming with mice begot of the mud of Nylus, to the great calamity of the inhabitants."

II. From Redi to Schwann.—The second period embraces the experimental tests of Redi (1668), Spallanzani (1775), and Schwann (1837)—notable achievements that resulted in a verdict for the adherents to the doctrine of biogenesis. Here the question might have rested had it not been opened upon theoretical ground by Pouchet in 1859.

The First Experiments.—The belief in spontaneous generation, which was so firmly implanted in the minds of naturalists, was subjected to an experimental test in 1668 by the Italian Redi. It is a curious circumstance, but one that throws great light upon the condition of intellectual development of the period, that no one previous to Redi had attempted to test the truth or falsity of the theory of spontaneous generation. To approach this question from the experimental side was to do a great service to science.

The experiments of Redi were simple and homely. He exposed meat in jars, some of which were left uncovered, some covered with parchment, and others with fine wire gauze. The meat in all these vessels became spoiled, and flies, being attracted by the smell of decaying meat, laid eggs in that which was exposed, and there came from it a large crop of maggots. The meat which was covered by parchment also decayed in a similar manner, without the appearance of maggots within it; and in those vessels covered by wire netting the flies laid their eggs upon the wire netting. There they hatched, and the maggots, instead of appearing in the meat, appeared on the surface of the wire gauze. From this Redi concluded that maggots arise in decaying meat from the hatching of the eggs of insects, but inasmuch as these animals had been supposed to arise spontaneously within the decaying meat, the experiment took the ground from under that hypothesis.

He made other observations on the generation of insects, but with acute scientific analysis never allowed his conclusions to run ahead of his observations. He suggested, however, the probability that all cases of the supposed production of life from dead matter were due to the introduction of living germs from without. The good work begun by Redi was confirmed and extended by Swammerdam (1637–1681) and Vallisnieri (1661–1730), until the notion of the spontaneous origin of any forms of life visible to the unaided eye was banished from the minds of scientific men.

Redi (Fig. 89) was an Italian physician living in Arentino, distinguished alike for his attainments in literature and for his achievements in natural science. He was medical adviser to two of the grand dukes of Tuscany, and a member of the Academy of Crusca. Poetry as well as other literary compositions shared his time with scientific occupations. His collected works, literary, scientific, and medical, were published in nine octavo volumes in Milan, 1809–1811. This collection includes his life and letters, and embraces one

FIG. 89.—FRANCESCO REDI, 1626–1697.

volume of sonnets. The book that has been referred to as containing his experiments was entitled *Esperienze Intorno Alla Generazione Degl'Insetti*, and first saw the light in quarto form in Florence in 1668. It went through five editions in twenty years. Some of the volumes were translated into Latin, and were published in miniature, making books not more than four inches high. Huxley says: "The extreme simplicity of his experiments, and the clearness of his arguments, gained for his views and for their consequences almost universal acceptance."

New Form of the Question.—The question of the spontaneous generation of life was soon to take on a new aspect. Seven years after the experiments of Redi, Leeuwenhoek made known a new world of microscopic organisms—the infusoria—and, as we have seen, he discovered, in 1687, those still minuter forms, the bacteria. Strictly speaking, the bacteria, on account of their extreme minuteness, were lost sight of, but spontaneous generation was evoked to account for the birth of all microscopic organisms, and the question circled mainly around the infusorial animalcula. While the belief in the spontaneous generation of life among forms visible to the unaided eye had been surrendered, nevertheless doubts were entertained as to the origin of microscopic organisms, and it was now asserted that here were found the beginnings of life—the place where inorganic material was changed through natural agencies into organized beings microscopic in size.

More than seventy years elapsed before the matter was again subjected to experimental tests. Then Needham, using the method of Redi, began to experiment on the production of microscopic animalcula. In many of his experiments he was associated with Buffon, the great French naturalist, who had a theory of organic molecules that he wished to sustain. Needham (1713–1784), a priest of the Catholic faith, was an Englishman living on the Continent; he was for many years director of the Academy of Maria Theresa at Brussels. He engaged in scientific investigations in connection with his work of teaching. The results of Needham's first experiments were published in 1748. These experiments were conducted by extracting the juices of meat by boiling; by then enclosing the juices in vials, the latter being carefully corked and sealed with mastic; by subjecting the sealed bottles, finally, to heat, and setting them away to cool. In due course of time, the fluids thus treated became infected with microscopic life, and, inasmuch as Needham believed that he had killed all living germs by repeated heating, he concluded that the living forms had been produced by spontaneous generation.

Spallanzani.—The epoch-making researches of Spallanzani, a fellow-countryman of Redi, were needed to point out the error in Needham's conclusions. Spallanzani (Fig. 90) was one of the most eminent men of his time. He was educated for the church, and, therefore, he is usually known under the title of Abbé Spallanzani. He did not, however, actively engage in his churchly offices, but, following an innate love of natural science and of investigation, devoted himself to experiments and researches and to teaching. He was first a professor at Bologna, and afterward at the University of Pavia. He made many additions to knowledge of the development and the physiology of organisms, and he was the first to make use of glass flasks in the experimental study of the question of the spontaneous generation of life.

Spallanzani thought that the experiments of Needham had not been conducted with sufficient care and precision; accordingly, he made use of glass flasks with slender necks which could be hermetically sealed after the nutrient fluids had been introduced. The vials which Needham used as containers were simply corked and sealed with mastic, and it was by no means certain that the entrance of air after heating had been prevented; moreover, no record was made by Needham of the temperature and the time of heating to which his bottles and fluids had been subjected.

Spallanzani took nutrient fluids, such as the juices of vegetables and meats which had been extracted by boiling, placed them in clear flasks, the necks of which were hermetically sealed in flame, and afterward immersed them in boiling water for three-quarters of an hour, in order to destroy all

FIG. 90.—LAZZARO SPALLANZANI, 1729–1799.

germs that might be contained in them. The organic infusions of Spallanzani remained free from change. It was then, as now, a well-known fact that organic fluids, when exposed to air, quickly decompose and acquire a bad smell;

they soon become turbid, and in a little time a scum is formed upon their surface. The fluids in the flasks of Spallanzani remained of the same appearance and consistency as when they were first introduced into the vessel, and the obvious conclusion was drawn that microscopic life is not spontaneously formed within nutrient fluids.

"But Needham was not satisfied with these results, and with a show of reason maintained that such a prolonged boiling would destroy not only germs, but the germinative, or, as he called it, the 'vegetative force' of the infusion itself. Spallanzani easily disposed of this objection by showing that when the infusions were again exposed to the air, no matter how severe or prolonged the boiling to which they had been subjected, the infusoria reappeared. His experiments were made in great numbers, with different infusions, and were conducted with the utmost care and precision" (Dunster). It must be confessed, however, that the success of his experiments was owing largely to the purity of the air in which he worked, the more resistant atmospheric germs were not present: as Wyman showed, long afterward, that germs may retain their vitality after being subjected for several hours to the temperature of boiling water.

Schulze and Schwann.—The results of Spallanzani's experiments were published in 1775, and were generally regarded by the naturalists of that period as answering in the negative the question of the spontaneous generation of life. Doubts began to arise as to the conclusive nature of Spallanzani's experiments, on account of the discovery of the part which oxygen plays in reference to life. The discovery of oxygen, one of the greatest scientific events of the eighteenth century, was made by Priestley in 1774. It was soon shown that oxygen is necessary to all forms of life, and the question was raised: Had not the boiling of the closed flasks changed the oxygen so that through the heating process it had lost its life-giving properties? This doubt grew until a reëxamination of the question of spontaneous generation became necessary under conditions in which the nutrient fluids were made accessible to the outside air.

In 1836 Franz Schulze, and, in the following year, Theodor Schwann, devised experiments to test the question on this new basis. Schwann is known to us as the founder of the cell-theory, but we must not confuse Schulze with Max Schultze, who established the protoplasm doctrine. In the experiments of Schulze, a flask was arranged containing nutrient fluids, with a large cork perforated and closely fitted with bent glass tubes connected on one side with a series of bulbs in which were placed sulphuric acid and other chemical substances. An aspirator was attached to the other end of this system, and air from the outside was sucked into the flask, passing on its way through the bulbs containing the chemical substances. The purpose of this was to remove the floating germs that exist in the air, while the air itself was shown, through other experiments by Schwann, to remain unchanged.

Tyndall says in reference to these experiments: "Here again the success of Schulze was due to his working in comparatively pure air, but even in such air his experiment is a risky one. Germs will pass unwetted and unscathed through sulphuric acid unless the most special care is taken to detain them. I have repeatedly failed, by repeating Schulze's experiments, to obtain his results. Others have failed likewise. The air passes in bubbles through the bulbs, and to render the method secure, the passage of the air must be so slow as to cause the whole of its floating matter, even to the very core of each bubble, to touch the surrounding fluid. But if this precaution be observed *water will be found quite as effectual as sulphuric acid.*"

Schwann's apparatus was similar in construction, except

that the bent tube on one side was surrounded by a jacket of metal and was subjected to a very high temperature while the air was being drawn through it, the effect being to kill any floating germs that might exist in the air. Great care was taken by both experimenters to have their flasks and fluids thoroughly sterilized, and the results of their experiments were to show that the nutrient fluids remained uncontaminated.

These experiments proved that there is something in the atmosphere which, unless it be removed or rendered inactive, produces life within nutrient fluids, but whether this something is solid, fluid, or gaseous did not appear from the experiments. It remained for Helmholtz to show, as he did in 1843, that this something will not pass through a moist animal membrane, and is therefore a solid. The results so far reached satisfied the minds of scientific men, and the question of the spontaneous origin of life was regarded as having been finally set at rest.

III. The Third Period. Pouchet.—We come now to consider the third historical phase of this question. Although it had apparently been set at rest, the question was unexpectedly opened again in 1859 by the Frenchman Pouchet, the director of the Natural History Museum of Rouen. The frame of mind which Pouchet brought to his experimental investigations was fatal to unbiased conclusions: "When, *by meditation*," he says, in the opening paragraph of his book on *Heterogenesis*, "it was evident to me that spontaneous generation was one of the means employed by nature for the production of living beings, I applied myself to discover by what means one could place these phenomena in evidence." Although he experimented, his case was prejudiced by metaphysical considerations. He repeated the experiments of previous observers with opposite results, and therefore he declared his belief in the falsity of the conclusions of Spallanzani, Schulze, and Schwann.

He planned and executed one experiment which he supposed was conclusive. In introducing it he said: "The opponents of spontaneous generation assert that the germs of microscopic organisms exist in the air, which transports them to a distance. What, then, will these opponents say if I succeed in introducing the generation of living organisms, while substituting artificial air for that of the atmosphere?"

He filled a flask with boiling water and sealed it with great care. This he inverted over a bath of mercury, thrusting the neck of the bottle into the mercury. When the water was cooled, he opened the neck of the bottle, still under the mercury, and connected it with a chemical retort containing the constituents for the liberation of oxygen. By heating the retort, oxygen was driven off from the chemical salts contained in it, and being a gas, the oxygen passed through the connecting tube and bubbled up through the water of the bottle, accumulating at the upper surface, and by pressure forcing water out of the bottle. After the bottle was about half filled with oxygen imprisoned above the water, Pouchet took a pinch of hay that had been heated to a high temperature in an oven, and with a pair of sterilized forceps pushed it underneath the mercury and into the mouth of the bottle, where the hay floated into the water and distributed itself.

He thus produced a hay infusion in contact with pure oxygen, and after a few days this hay infusion was seen to be cloudy and turbid. It was, in fact, swarming with micro-organisms. Pouchet pointed with triumphant spirit to the apparently rigorous way in which his experiment had been carried on: "Where," said he, "does this life come from? It can not come from the water which had been boiled, destroying all living germs that may have existed in it. It can not come from the oxygen which was produced at the temperature of incandescence. It can not have been carried in the hay, which had been heated for a long period before being introduced into the water." He declared that this life was, therefore, of spontaneous origin.

The controversy now revived, and waxed warm under the insistence of Pouchet and his adherents. Finally the Academy of Sciences, in the hope of bringing it to a conclusion, appointed a committee to decide upon conflicting claims.

Pasteur.—Pasteur had entered into the investigation of the subject about 1860, and, with wonderful skill and acumen, was removing all possible grounds for the conclusions of Pouchet and his followers. In 1864, before a brilliant audience at the Sorbonne, he repeated the experiment outlined above and showed the source of error. In a darkened room he directed a bright beam of light upon the apparatus, and his auditors could see in the intense illumination that the surface of the mercury was covered with dust particles. Pasteur then showed that when a body was plunged beneath the mercury, some of these surface granules were carried with it. In this striking manner Pasteur demonstrated that particles from the outside had been introduced into the bottle of water by Pouchet. This, however, is probably not the only source of the organisms which were developed in Pouchet's infusions. It is now known that a hay infusion is very difficult to sterilize by heat, and it is altogether likely that the infusions used by Pouchet were not completely sterilized.

The investigation of the question requires more critical methods than was at first supposed, and more factors enter into its solution than were realized by Spallanzani and Schwann.

Pasteur demonstrated that the floating particles of the air contained living germs, by catching them in the meshes of gun cotton, and then dissolving the cotton with ether and examining the residue. He also showed that sterilized organic fluids could be protected by a plug of cotton sufficiently porous to admit of exchange of air, but matted closely enough to entangle the floating particles. He showed also that many of the minute organisms do not require free oxygen for their life processes, but are able to take the oxygen by chemical decomposition which they themselves produce from the nutrient fluids.

Jeffries Wyman, of Harvard College, demonstrated that some germs are so resistant to heat that they retain their vitality after several hours of boiling. This fact probably accounts for the difference in the results that have been obtained by experimenters. The germs in a resting-stage are surrounded by a thick protective coat of cellulose, which becomes softened and broken when they germinate. On this account more recent experimenters have adopted a method of discontinuous heating of the nutrient fluid that is being tested. The fluids are boiled at intervals, so that the unusually resistant germs are killed after the coating has been rendered soft, and when they are about to germinate.

After the brilliant researches of Pasteur, the question of spontaneous germination was once again regarded as having been answered in the negative; and so it is regarded to-day by the scientific world. Nevertheless, attempts have been made from time to time, as bv Bastian, of England, in 1872, to revive it on the old lines.

Tyndall.—John Tyndall (1820–1893), the distinguished physicist, of London, published, in 1876, the results of his experiments on this question, which, for clearness and ingenuity, have never been surpassed. For some time he had been experimenting in the domain of physics with what he called optically pure air. It was necessary for him to have air from which the floating particles had been sifted, and it occurred to him that he might expose nutrient fluids to this optically pure air, and thus very nicely test the question of the spontaneous origin of life within them.

He devised a box, or chamber, as shown in Fig. 91, having in front a large glass window, two small glass windows on the ends, and in the back a little air-tight trap-door. Through the bottom of this box he had fitted ordinary test

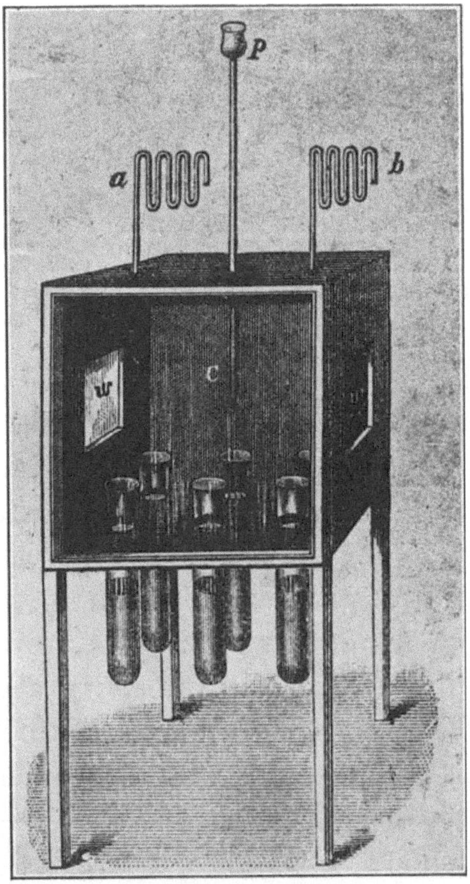

FIG. 91.—Apparatus of Tyndall for Experimenting on Spontaneous Generation.

tubes of the chemist, with an air-tight surrounding, and on the top he had inserted some coiled glass tubes, which were open at both ends and allowed the passage of air in and out of the box through the tortuous passage. In the middle of the top of the box was a round piece of rubber. When he perforated this with a pinhole the elasticity of the rubber would close the hole again, but it would also admit of the passage through it of a small glass tube, such as is called by chemists a "thistle tube." The interior of this box was painted with a sticky substance like glycerin, in order to retain the floating particles of the air when they had once settled upon its sides and bottom. The apparatus having been prepared in this way, was allowed to stand, and the floating particles settled by their own weight upon the bottom and sides of the box, so that day by day the number of floating particles became reduced, and finally all of them came to rest.

The air now differed from the outside air in having been purified of all of its floating particles. In order to test the complete disappearance of all particles. Tyndall threw a beam of light into the air chamber. He kept his eye in the darkness for some time in order to increase its sensitiveness; then, looking from the front through the glass into the box, he was able to see any particles that might be floating there. The floating particles would be brightly illuminated by the condensed light that he directed into the chamber, and would become visible. When there was complete darkness within the chamber, the course of the beam of light was apparent in the room as it came up to the box and as it left the box, being seen on account of the reflection from the floating particles in the air, but it could not be seen at all within the box. When this condition was reached, Tyndall had what he called optically pure air, and he was now ready to introduce the nutrient fluids into his test tubes. Through a thistle tube, thrust into the rubber diaphragm above, he was able to bring the mouth of the tube successively over the different test tubes, and, by pouring different kinds of fluids from above, he was able to introduce these into different test tubes. These fluids consisted of mutton broth, of turnip-broth, and other decoctions of animal and vegetable matter. It is to be noted that the test tubes were not corked and consequently that the fluids contained within them were freely exposed to the optically pure air within the chamber.

The box was now lifted, and the ends of the tubes extending below it were thrust into a bath of boiling oil. This set the fluids into a state of boiling, the purpose being to kill any germs of life that might be accidentally introduced into them in the course of their conveyance to the test tubes. These fluids, exposed freely to the optically pure air within this chamber, then remained indefinitelv free from micro-organisms, thus demonstrating that putrescible fluids may be freely exposed to air from which the floating particles have been removed, and not show a trace either of spoiling or of organic life within them.

It might be objected that the continued boiling of the fluids had produced chemical changes inimical to life, or in some way destroyed their life-supporting properties; but after they had remained for months in a perfectly clear state, Tyndall opened the little door in the back of the box and closed it at once, thereby admitting some of the floating particles from the outside air. Within a few days' time the fluids which previously had remained uncontaminated were spoiling and teeming with living organisms.

These experiments showed that under the conditions of the experiments no spontaneous origin of life takes place. But while we must regard the hypothesis of spontaneous generation as thus having been disproved on an experimental basis, it is still adhered to from the theoretical standpoint by many naturalists; and there are also many who think that life arises spontaneously at the present time in ultra-microscopic particles. Weismann's hypothetical "biophors," too minute for microscopic observation, are supposed to arise by spontaneous generation. This phase of the question, however, not being amenable to scientific tests, is theoretical, and therefore, so far as the evidence goes, we may safely say that the spontaneous origin of life under present conditions is unknown.

Practical Applications.—There are, of course, numerous practical applications of the discovery that the spoiling of putrescible fluids is due to floating germs that have been introduced from the air. One illustration is the canning of meats and fruits, where the object is, by heating, to destroy all living germs that are distributed through the substance, and then, by canning, to keep them out. When this is entirely successful, the preserved vegetables and meats go uncontaminated. One of the most important and practical applications came in the recognition (1867) by the English surgeon Lister that wounds during surgical operations are poisoned by floating particles in the air or by germs clinging to instruments or the skin of the operator, and that to render all appliances sterile and, by antiseptic dressings, completely to prevent the entrance of these bacteria into surgical wounds, insures their being clean and healthy. This led to antiseptic surgery, with which the name of Lister is indissolubly connected.

The Germ-Theory of Disease

The germ-theory of disease is another question of general bearing, and it will be dealt with briefly here.

After the discovery of bacteria by Leeuwenhoek, in 1687, some medical men of the time suggested the theory that contagious diseases were due to microscopic forms of life that passed from the sick to the well. This doctrine of *contagium vivum*, when first promulgated, took no firm root, and gradually disappeared. It was not revived until about 1840. If we attempt briefly to sketch the rise of the germ-theory of disease, we come, then, first to the year 1837, when the Italian Bassi investigated the disease of silkworms, and showed that the transmission of that disease was the result of the passing of minute glittering particles from the sick to the healthy. Upon the basis of Bassi's observation, the distinguished anatomist Henle, in 1840, expounded the theory that all contagious diseases are due to microscopic germs.

The matter, however, did not receive experimental proof until 1877, when Pasteur and Robert Koch showed the direct connection between certain microscopic filaments and the disease of splenic fever, which attacks sheep and other cattle. Koch was able to get some of these minute filaments under the microscope, and to trace upon a warm stage the different steps in their germination. He saw the spores bud and produce filamentous forms. He was able to cultivate these upon a nutrient substance, gelatin, and in this way to obtain a pure culture of the organism, which is designated under the term anthrax. He inoculated mice with the pure culture of anthrax germs, and produced splenic fever in the inoculated forms. He was able to do this through several generations of mice. In the same year Pasteur showed a similar connection between splenic fever and the anthrax.

This demonstration of the actual connection between anthrax and splenic fever formed the first secure foundation of the germ-theory of disease, and this department of investigation became an important one in general biology. The pioneer workers who reached the highest position in the development of this knowledge are Pasteur, Koch, and Lister.

Veneration of Pasteur.—Pasteur is one of the most conspicuous figures of the nineteenth century. The veneration in which he is held by the French people is shown in the result of a popular vote, taken in 1907, by which he was placed at the head of all their notable men. One of the most

Fig. 92.—Louis Pasteur (1822–1895) and his Granddaughter.

widely circulated of the French journals—the *Petit Parisien*—appealed to its readers all over the country to vote upon the relative prominence of great Frenchmen of the last century. Pasteur was the winner of this interesting contest, having received 1,338,425 votes of the fifteen millions cast, and ranking above Victor Hugo, who stood second in popular estimation, by more than one hundred thousand votes. This enviable recognition was won, not by spectacular achievements in arms or in politics, but by indefatigable industry in the quiet pursuit of those scientific researches that have resulted in so much good to the human race.

Personal Qualities.—He should be known also from the side of his human qualities. He was devotedly attached to his family, enjoying the close sympathy and assistance of his wife and his daughter in his scientific struggles, a circumstance that aided much in ameliorating the severity of his labors. His labors, indeed, overstrained his powers, so that he was smitten by paralysis in 1868, at the age of forty-six, but with splendid courage he overcame this handicap, and continued his unremitting work until his death in 1895.

The portrait of Pasteur with his granddaughter (Fig. 92) gives a touch of personal interest to the investigator and the contestant upon the field of science. His strong face shows dignity of purpose and the grim determination which led to colossal attainments; at the same time it is mellowed by gentle affection, and contrasts finely with the trusting expression of the younger face.

Pasteur was born of humble parents in Dôle in the Jura, on December the 27th, 1822. His father was a tanner, but withal, a man of fine character and stern experience, as is "shown by the fact that he had fought in the legions of the First Empire and been decorated on the field of battle by Napoleon." The filial devotion of Pasteur and his justifiable pride in his father's military service are shown in the dedication of his book, *Studies on Fermentation*, published in 1876:

"To the memory of my Father,

Formerly a soldier under the First Empire, and Knight of the Legion of Honor.

The longer I live, the better do I understand the kindness of thy heart and the superiority of thy judgment.

The efforts which I have devoted to these studies and to those which have preceded them are the fruits of thy example and of thy counsel.

Desiring to honor these precious recollections, I dedicate this book to thy memory."

When Pasteur was an infant of two years his parents removed to the town of Arbois, and here he spent his youth and received his early education. After a period of indifference to study, during which he employed his time chiefly in fishing and sketching, he settled down to work, and, thereafter, showed boundless energy and enthusiasm.

Pasteur, whom we are to consider as a biologist, won his first scientific recognition at the age of twenty-five, in chemistry and molecular physics. He showed that crystals of certain tartrates, identical in chemical composition, acted differently upon polarized light transmitted through them. He concluded that the differences in optical properties depended upon a different arrangement of the molecules; and these studies opened the fascinating field of molecular physics and physical chemistry.

Pasteur might have remained in this field of investigation, but his destiny was different. As Tyndall remarked, "In the investigation of microscopic organisms—the 'infinitely little,' as Pasteur loved to call them—and their doings in this, our world, Pasteur found his true vocation. In this broad field it has been his good fortune to alight upon a crowd of connected problems of the highest public and scientific interest, ripe for solution, and requiring for their successful treatment the precise culture and capacities which he has brought to bear upon them."

In 1857 Pasteur went to Paris as director of scientific studies in the École Normale, having previously been a professor in Strasburg and in Lille. From this time on his energies became more and more absorbed in problems of a biological nature. It was a momentous year (1857) in the annals of bacteriology when Pasteur brought convincing proof that fermentation (then considered chemical in its nature) was due to the growth of organic life. Again in 1860 he demonstrated that both lactic (the souring of milk) and alcoholic fermentation are due to the growth of microscopic organisms, and by these researches he developed the province of biology that has expanded into the science of bacteriology.

After Pasteur entered the path of investigation of microbes his progress was by ascending steps; each new problem the solution of which he undertook seemed of greater importance than the one just conquered. He was led from the discovery of microbe action to the application of his knowledge to the production of antitoxins. In all this he did not follow his own inclinations so much as his sense of a call to service. In fact, he always retained a regret that he was not permitted to perfect his researches on crystallography. At the age of seventy he said of himself: "If I have a regret, it is that I did not follow that route, less rude it seems to me, and which would have led, I am convinced, to wonderful discoveries. A sudden turn threw me into the study of fermentation, fermentations set me at diseases, but I am still inconsolable to think that I have never had the time to go back to my old subject" (Tarbell).

Although the results of his combined researches form a succession of triumphs, every point of his doctrines was the subject of fierce controversy; no investigations ever met with more determined opposition, no investigator ever fought more strenuously for the establishment of each new truth.

He went from the study of the diseases of wines (1865) to the investigation (1865–1868) of the silkworm plague which had well-nigh crushed the silk industry of his country. The result was the saving of millions of francs annually to the people of France.

His Supreme Service.—He then entered upon his chief services to humanity—the application of his discoveries to the cure and prevention of diseases. By making a succession of pure cultures of a disease-producing virus, he was able to attenuate it to any desired degree, and thereby to create a vaccinating form of the virus capable of causing a mild affection of the disease. The injection of this attenuated virus secured immunity from future attacks. The efficacy of this form of inoculation was first proved for the disease of fowl cholera, and then came the clear demonstration (1881) that the vaccine was effective against the splenic fever of cattle. Crowning this series of discoveries came the use of inoculation (1885) to prevent the development of hydrophobia in one bitten by a mad dog.

The Pasteur Institute.—The time had now come for the establishment of an institute, not alone for the treatment of hydrophobia, but also for the scientific study of means to control other diseases, as diphtheria, typhoid, tuberculosis, etc. A movement was set on foot for a popular subscription to meet this need. The response to this call on the part of the common people was gratifying. "The extraordinary enthusiasm which accompanied the foundation of this great institution has certainly not been equaled in our time. Considerable sums of money were subscribed in foreign countries, while contributions poured in from every part of France. Even the inhabitants of obscure little towns and villages organized fêtes, and clubbed together to send their small gifts" (Francklаnd). The total sum subscribed on the date of the opening ceremony amounted to 3,586,680 francs.

The institute was formally opened on November 14th, 1888, with impressive ceremonies presided over by the President of the Republic of France. The establishment of this institute was an event of great scientific importance. Here, within the first decade of its existence, were successfully treated more than twenty thousand cases of hydrophobia. Here has been discovered by Roux the antitoxin for diphtheria, and here have been established the principles of inoculation against the bubonic plague, against lockjaw, against tuberculosis and other maladies, and of the recent microbe inoculations of Wright of London. More than thirty "Pasteur institutes," with aims similar to the parent institution, have been established in different parts of the civilized world.

Pasteur died in 1895, greatly honored by the whole world. On Saturday, October 5th of that year, a national funeral was conducted in the Church of Notre-Dame, which was attended by the representatives of the state and of numerous scientific bodies and learned societies.

Koch.—Robert Koch (Fig. 93) was born in 1843, and is still living, engaged actively in work in the University of Berlin. His studies have been mainly those of a medical man, and have been crowned with remarkable success. In 1881 he discovered the germ of tuberculosis, in 1883 the germ that produces Asiatic cholera, and since that time his name has been connected with a number of remarkable discoveries that are of continuous practical application in the science of medicine.

Koch, with the rigorous scientific spirit for which he is noteworthy, established four necessary links in the chain of evidence to show that a particular organism is connected with a particular disease. These four postulates of Koch are:

First, that a microscopic organism of a particular type should be found in great abundance in the blood and the tissue of the sick animal; second, that a pure culture should be made of the suspected organism; third, that this pure culture, when introduced into the body of another animal, should produce

FIG. 93.—ROBERT KOCH, BORN 1843.

the disease; and, fourth, that in the blood and tissues of that animal there should be found quantities of the particular organism that is suspected of producing the disease. In the case of some diseases this entire chain of evidence has been established; but in others, such as cholera and typhoid fever, the last steps have not been completed, for the reason that the animals experimented upon, namely, guinea-pigs, rabbits, and mice, are not susceptible to these diseases.

Lister.—The other member of the great triumvirate of bacteriology is Sir Joseph Lister (Fig. 94); born in 1827, he has been successively professor of surgery in the universities

FIG. 94.—SIR JOSEPH LISTER, BORN 1827.

of Glasgow (1860) and of Edinburgh (1869), and in King's College, London (1877). His practical application of the germ-theory introduced aseptic methods into surgery and completely revolutionized that field. This was in 1867. In an address given that year before the British Medical Association in Dublin, he said: "When it had been shown by the researches of Pasteur that the septic property of the atmosphere depended, not on oxygen or any gaseous constituent, but on minute organisms suspended in it, which owed their energy to their vitality, it occurred to me that decomposition in the injured part might be avoided without excluding the air, by applying as a dressing some material capable of destroying the life of the floating particles." At first he used carbolic acid for this purpose. "The wards of which he had charge in the Glasgow Infirmary were especially affected by gangrene, but in a short time became the healthiest in the world; while other wards separated by a passageway retained their infection." The method of Lister has been universally adopted, and at the same time has been greatly extended and improved.

The question of immunity, *i.e.*, the reason whv after having had certain contagious diseases one is rendered immune, is of very great interest, but is of medical bearing, and therefore is not dealt with here.

Bacteria and Nitrates.—One further illustration of the connection between bacteria and practical affairs may be mentioned. It is well known that animals are dependent upon plants, and that plants in the manufacture of protoplasm make use of certain nitrites and nitrates which they obtain from the soil. Now, the source of these nitrites and nitrates is very interesting. In animals the final products of broken-down protoplasm are carbon dioxide, water, and a nitrogenous substance called urea. These products are called excretory products. The animal machine is unable to utilize the energy which exists in the form of potential energy in these substances, and they are removed from the body

The history of nitrogenous substance is the one which at present interests us the most. Entering the soil, it is there acted upon by bacteria residing in the soil, these bacteria possessing the power of making use of the lowest residuum of energy left in the nitrogenous substance. They cause the nitrogen and the hydrogen to unite with oxygen in such a way that there are produced nitrous and nitric acids, and from these two acids, through chemical action, result the nitrites and the nitrates. These substances are then utilized by the plant in the manufacture of protoplasm, and the plant is fed upon by animal organisms, so that a direct relationship is established between these lower forms of life and the higher plant and animal series; a relationship that is not only interesting, but that helps to throw an important side-light upon the general nature of vital activities, their kind and their reach. In addition to the soil bacteria mentioned above, there are others that form association with the rootlets of certain plants and possess the power of fixing free nitrogen from the air.

The nitrifying bacteria, are, of course, of great importance to the farmer and the agriculturist.

It is not our purpose, however, to trace the different phases of the subject of bacteriology to their conclusions, but rather to give a picture of the historical development of this subject as related to the broader one of general biology.

CHAPTER XIV

HEREDITY AND GERMINAL CONTINUITY—MENDEL, GALTON, WEISMANN

It is a matter of common observation that in the living world like tends to produce like. The offspring of plants, as well as of animals, resembles the parent, and among all organisms endowed with mind, the mental as well as the physical qualities are inherited. This is a simple statement of the fact of heredity, but the scientific study of inheritance involves deep-seated biological questions that emerged late in the nineteenth century, and the subject is still in its infancy.

In investigating this question, we need first, if possible, to locate the bearers of hereditary qualities within the physical substance that connects one generation with the next; then, to study their behavior during the transmission of life in order to account for the inheritance of both maternal and paternal qualities; and, lastly, to determine whether or not transiently acquired characteristics are inherited.

Hereditary Qualities in the Germinal Elements.—When we take into consideration the fact established for all animals and plants (setting aside cases of budding and the division of unicellular organisms), that the only substance that passes from one generation to another is the egg and the sperm in animals, and their representatives in plants, we see that the first question is narrowed to these bodies. If all hereditary qualities are carried in the egg and the sperm—as it seems they must be—then it follows that these germinal elements, although microscopic in size, have a very complex organization. The discovery of this organization must depend upon microscopic examination. Knowledge regarding the physical basis of heredity has been greatly advanced by critical studies of cells under the microscope and by the application of experimental methods, while other phases of the problems of inheritance have been elucidated by the analysis of statistics regarding hereditary transmissions. The whole question, however, is so recent that a clear formulation of the direction of the main currents of progress will be more helpful than any attempt to estimate critically the underlying principles.

Early Theories.—There were speculations regarding the nature of inheritance in ancient and mediæval times. To mention any of them prior to the eighteenth century would serve no useful purpose, since they were vague and did not form the foundation upon which the modern theories were built. The controversies over pre-formation and epigenesis (see Chapter X) of the eighteenth century embodied some ideas that have been revived. The recent conclusion that there is in the germinal elements an inherited organization of great complexity which conditions inheritance seems, at first, to be a return to the doctrine of pre-formation, but closer examination shows that there is merely a general resemblance between the ideas expressed by Haller, Bonnet, and philosophers of their time and those current at the present time. Inherited organization, as now understood, is founded on the idea of germinal continuity and is vastly different from the old theory of pre-formation. The meaning of epigenesis, as expressed by Wolff, has also been modified to include the conception of pre-localization of hereditary qualities within particular parts of the egg. It has come now to mean that development is a process of differentiation of certain qualities already laid down in the germinal elements.

Darwin's Theory of Pangenesis.—In attempting to

account for heredity, Darwin saw clearly the necessity of providing some means of getting all hereditary qualities combined within the egg and the sperm. Accordingly he originated his provisional theory of pangenesis. Keeping in mind the fact that all organisms begin their lives in the condition of single cells, the idea of inheritance through these microscopic particles becomes difficult to understand. How is it possible to conceive of all the hereditary qualities being contained within the microscopic germ of the future being? Darwin supposed that very minute particles, which he called gemmules, were set free from all the cells in the body, those of the muscular system, of the nervous system, of the bony tissues, and of all other tissues contributing their part. These liberated gemmules were supposed to be carried by the circulation and ultimately to be aggregated within the germinal elements (ovum and sperm). Thus the germinal elements would be a composite of substances derived from all organs and all tissues.

With this conception of the blending of the parental qualities within the germinal elements we can conceive how inheritance would be possible and how there might be included in the egg and the sperm a representative in material substance of all the qualities of the parents. Since development begins in a fertilized ovum, this complex would contain minute particles derived from every part of the bodies of both parents, which by growth would give rise to new tissues, all of them containing representatives of the tissues of the parent form.

Theory of Pangenesis Replaced by that of Germinal Continuity.—This theory of Darwin served as the basis for other theories founded upon the conception of the existence of pan gens; and although the modifications of Spencer, Brooks, and others were important, it is not necessary to indicate them in detail in order to understand what is to follow. The various theories founded upon the idea of pangens were destined to be replaced by others founded on the conception of geminal continuity—the central idea in nineteenth-century biology.

The four chief steps which have led to the advancement of the knowledge of heredity, as suggested by Thomson, are as follows: " (a) The exposition of the doctrine of germinal continuity. (b) More precise investigation of the material basis of inheritance. (c) Suspicions regarding the inheritance of acquired characteristics. (d) Application of statistical methods which have led to the formulation of the law of ancestral heredity " We shall take these up in order.

Exposition of the Doctrine of Germinal Continuity.—From parent to offspring there passes some hereditary substance; although small in amount, it is the only living thread that connects one generation with another. It thus appears that there enters into the building of the body of a new organism some of the actual substance of both parents, and that this transmitted substance must be the bearer of hereditary qualities. Does it also contain some characteristics inherited from grandparents and previous generations? If so, how far back in the history of the race does unbroken continuity extend?

Briefly stated, genetic continuity means that the ovum and its fertilizing agent are derived by continuous cell-lineage from the fertilized ovum of previous generations, extending back to the beginning of life. The first clear exposition of this theorv occurs in the classical work of Virchow on *Cellular Pathology*, published in 1858. Virchow (1821–1902), the distinguished professor of the University of Berlin, has already been spoken of in connection with the development of histology. He took the step of overthrowing the theory of free cell-formation, and replacing it by the doctrine of cell-succession. According to the theory of Schleiden and Schwann, cells arose from a blastema by a condensation of matter around a nucleus, and the medical men prior to 1858 believed in free cell-formation within a matrix of secreted or excreted substance. This doctrine was held with tenacity especially for pathological growths. Virchow demonstrated, however, that there is a continuity of living substance in all growths—that cells, both in health and in disease, arise only by the growth and division of previously existing living cells; and to express this truth he coined the formula "*omnis cellula e cellula*." Manifestly it was necessary to establish this law of cell-succession before any idea of germinal continuity could prevail. Virchow's work in this connection is of undying value.

When applied to inheritance the idea of the continuity of living substance leads to making a distinction between germ-cells and body-cells. This had been done before the observations of Virchow made their separation of great theoretical value. Richard Owen, in 1849, pointed out certain differences between the body-cells and the germinal elements, but he did not follow up the distinction which he made. Haeckel's *General Morphology*, published in 1866, forecasts the idea also, and in 1878 Jaeger made use of the phrase "continuity of the germ protoplasm." Other suggestions and modifications led to the clear expression by Nussbaum, about 1875, that the germinal substance was continued by unbroken generations from the past, and is the particular substance in which all hereditary qualities are included. But the conception finds its fullest expression in the work of Weismann.

Weismann's explanation of heredity is at first sight relatively simple. In reply to the question, "Why is the offspring like the parent?" he says, "Because it is composed of some of the same stuff." In other words, there has been unbroken germinal continuity between generations. His idea of germinal continuity, *i.e.*, unbroken continuity, through all time, of the germinal substance, is a conception of very great extent, and now underlies all discussion of heredity.

In order to comprehend it, we must first distinguish between the germ-cells and the body-cells. Weismann regards the body, composed of its many cells, as a derivative that becomes simply a vehicle for the germ-cells. Owen's distinction between germ-cells and body-cells, made in 1849, was not of much importance, but in the theory of Weismann it is of vital significance. The germ-cells are the particular ones which carry forward from generation to generation the life of the individual. The body-cells are not inherited directly, but in the transmission of life the germ-cells pass to the succeeding generation, and they in turn have been inherited from the previous generation, and, therefore, we have the phenomenon of an unbroken connection with all previous generations.

When the full significance of this conception comes to us, we see why the germ-cells have an inherited organization of remarkable complexity. This germinal substance embodies all the past history of the living, impressionable protoplasm, which has had an unbroken series of generations. During all time it has been subjected to the molding influence of external circumstances to which it has responded, so that the summation of its experiences becomes in some way embedded within its material substance. Thus we have the germinal elements possessing an inherited organization made up of all the previous experiences of the protoplasm, some of which naturally are much more dominant than the others.

We have seen that this idea was not first expressed by Weismann; it was a modification of the views of Nussbaum and Hertwig. While it was not his individually, his conclusions were apparently reached independently. This idea was in the intellectual atmosphere of the times. Several investigators reached their conclusions independently, although there is great similarity between them. Although the credit for the first formulation of the law of germinal continuity does not belong to Weismann, that of the greatest elaboration of it does. This doctrine of germinal continuity is now so firmly embedded in biological ideas of inheritance and the evolution of animal life that we may say it has become the corner-stone of modern biology.

The conclusion reached—that the hereditary substance is the germ-plasm—is merely preliminary; the question remains, Is the germ-plasm homogeneous and endowed equally in all parts with a mixture of hereditary qualities? This leads to the second step.

The More Precise Investigation of the Material Basis of Inheritance.—The application of the microscope to critical studies of the structure of the germ-plasm has brought important results which merge with the development of the idea of germinal continuity. Can we by actual observation determine the particular part of the protoplasmic substance that carries the hereditary qualities? The earliest answer to this question was that the protoplasm, being the living substance, was the bearer of heredity. But close analysis of the behavior of the nucleus during development led, about 1875, to the idea that the hereditary qualities are located within the nucleus of the cell.

This idea, promulgated by Fol, Koelliker, and Oskar Hertwig, narrowed the attention of students of heredity from the general protoplasmic contents of the cell to the nucleus. Later investigations show that this restriction was, in a measure, right. The nucleus takes an active part during cell-division, and it was very natural to reach the conclusion that it is the particular bearer of hereditary substance. But, in 1883, Van Beneden and Boveri made the discovery that within the nucleus are certain distinct little rod-like bodies which make their appearance during cell-division. These little bodies, inasmuch as they stain very deeply with the dyes used in microscopic research, are called chromosomes. And continued investigation brought out the astounding fact that, although the number of chromosomes vary in different animals (commonly from two to twenty-four), they are of the same number in all the cells of any particular animal or plant. These chromosomes are regarded as the bearers of heredity, and their behavior during fertilization and development has been followed with great care.

Brilliant studies of the formation of the egg have shown that the egg nucleus, in the process of becoming mature, surrenders one-half its number of chromosomes; it approaches the surface of the egg and undergoes division, squeezing out one-half of its substance in the form of a polar globule; and this process is once repeated.* The formation of polar globules is accompanied by a noteworthy process of reduction in the number of chromosomes, so that when the egg nucleus has reached its mature condition it contains only one-half the number of chromosomes characteristic of the species, and will not ordinarily undergo development without fertilization.

The precise steps in the formation of the sperm have also been studied, and it has been determined that a parallel series of changes occur. The sperm, when it is fully formed, contains also one-half the number of chromosomes characteristic of the species. Now, egg and sperm are the two germinal elements which unite in development. Fertilization takes place by the union of sperm and egg, and inasmuch as the nuclei of each of these structures contain one-half of the number of chromosomes characteristic of the species,

* There are a few exceptions to this rule, as in the eggs of plant-lice, etc., in which a single polar globule is produced.

their union in fertilization results in the restoration of the original number of chromosomes. The fertilized ovum is the starting-point of a new organism, and from the method of its fertilization it appears that the parental qualities are passed along to the cells of every tissue.

The complex mechanism exhibited in the nucleus during segmentation is very wonderful. The fertilized ovum begins to divide, the nucleus passing through a series of complicated changes whereby its chromosomes undergo a lengthwise division—a division that secures an equable partition of the substance of which they are composed. With each successive division, this complicated process is repeated, and the many cells, arising from continued segmentation of the original cell, contain nuclei in which are embedded descendants of the chromosomes in unbroken succession. Moreover, since these chromosomes are bi-parental, we can readily understand that every cell in the body carries both maternal and paternal qualities.

The careful analysis of the various changes within the nuclei of the egg proves to be the key to some of the central questions of heredity. We see the force of the point which was made in a previous chapter, that inheritance is in the long run a cellular study, and we see in a new light the importance of the doctrine of germinal continuity. This conception, in fact, elucidates the general problem of inheritance in a way in which it has never been elucidated by any other means.

For some time the attention of investigators was concentrated upon the nucleus and the chromosomes, but it is now necessary to admit that the basis of some structures is discoverable within the cytoplasm that surrounds the nucleus. Experimental observations (Conklin, Lillie, Wilson) have shown the existence of particular areas within the apparently simple substance of the egg, areas which are definitely related to the development of particular parts of the embryo. The removal of any one of these pre-localized areas prevents the development of the part with which it is genetically related. Researches of this kind, necessitating great ingenuity in method and great talents in the observers, are widening the field of observation upon the phenomena of heredity.

The Inheritance of Acquired Characteristics.—The belief in the inheritance of acquired characteristics was generally accepted up to the middle of the nineteenth century, but the reaction against it started by Galton and others has assumed great proportions. Discussions in this line have been carried on extensively, and frequently in the spirit of great partizanship. These discussions cluster very much about the name and the work of Weismann, the man who has consistently stood against the idea of acquired characteristics. More in reference to this phase of the question is given in the chapter dealing with Weismann's theory of evolution (see p. 398). Wherever the truth may lie, the discussions regarding the inheritance of acquired characteristics provoked by Weismann's theoretical considerations, have resulted in stimulating experiment and research, and have, therefore, been beneficial to the advance of science.

The Application of Statistical Methods and Experiments to the Ideas of Heredity. Mendel.—This feature of investigating questions of heredity is of growing importance. The first to complete experiments and to investigate heredity to any purpose was the Austrian monk Mendel (1822–1884) (Fig. 95), the abbot of a monastery at Brünn. In his garden he made many experiments upon the inheritance, particularly in peas, of color and of form; and through these experiments he demonstrated a law of inheritance which bids fair to be one of the great biological discoveries of the nineteenth century. He published his papers in 1866 and 1867, but since the minds of naturalists at that time were very much occupied with the questions of organic evolution, raised through the publications of Darwin, the ideas of Mendel attracted very little attention. The principles that he established were re-discovered in 1900 bv De Vries and other botanists, and thus naturalists were led to look up the work of Mendel.

FIG. 95.—GREGOR MENDEL, 1822–1884.
Permlsslon of Professor Bateson.

The great discoverv of Mendel may be called that of the purity of the germ-cells. Bv cross-fertilization of pure breeds of peas of different colors and shapes he obtained hybrids. The hybrid embodied the characteristics of the crossed peas; one of the characteristics appearing, and the other being held in abeyance—present within the organization of the pea, but not visible. When peas of different color were cross-fertilized, one color would be stronger apparently than the other, and would stand out in the hybrids. This was called the dominant color. The other, which was held in abeyance, was called recessive; for, though unseen, it was still present within the young seeds. That the recessive color was not blotted out was clearly shown by raising a crop from the hybrid, a condition under which they would produce seeds like those of the two original forms, and in equal number; and thereafter the descendants of these peas would breed true. This so-called purity of the germ-cells, then, may be expressed in this way: "The hybrid, whatever its own character, produces ripe germ-cells, which produce only the pure character of one parent or of the other" (Castle).

Although Mendel's discovery was for a long time overlooked, happily the facts were re-discovered, and at the present time extensive experiments are being made with animals to test this law: experiments in the inheritance of poultry, the inheritance of fur in guinea-pigs, of erectness in the ears of rabbits, etc., etc. In this country the experiments of Castle, Davenport, and others with animals tend to support Mendel's conclusion and lift it to the position of a law.

Rank of Mendel's Discovery.—The discovery by Mendel of alternative inheritance will rank as one of the greatest discoveries in the study of heredity. The fact that in cross-breeding the parental qualities are not blended, but that they retain their individuality in the offspring, has many possible practical applications both in horticulture and in the breeding of animals. The germ-cells of the hybrids have the dominant and the recessive characters about equally divided; this will appear in the progeny of the second generation, and the races, when once separated, may be made to breed true.

Mendel's name was not recognized as a prominent one in the annals of biological history until the re-discovery of his law in 1900; but now he is accorded high rank. It may be remarked in passing that the three leading names in the development of the theories of heredity are those of Mendel, Galton, and Weismann.

Galton.—The application of statistical methods is well illustrated in the theories of Francis Galton (Fig. 96). This

FIG. 96.—FRANCIS GALTON, BORN 1822.

distinguished English statistician was born in 1822, and is still living. He is the grandson of Dr. Erasmus Darwin and the cousin of Charles. After publishing books on his travels in Africa, he began the experimental study of heredity and, in 1871, he read before the Royal Society of London a paper on Pangenesis, in which he departed from that theory as developed by Darwin. The observations upon which he based his conclusions were made upon the transfusion of blood in rabbits and their after-breeding. He studied the inheritance of stature, and other characteristics, in human families, and the inheritance of spots on the coat of certain hounds, and was led to formulate a law of ancestral inheritance which received its clearest expression in his book, *Natural Inheritance*, published in 1889.

He undertook to determine the proportion of heritage that is, on the average, contributed by each parent, grandparent, etc., and arrived at the following conclusions: "The parents together contribute one-half the total heritage, the four grandparents together one-fourth, the eight great-grandparents one-sixteenth, and all the remainder of the ancestry one-sixteenth."

Carl Pearson has investigated this law of ancestral inheritance. He substantiates the law in its principle, but modifies slightly the mathematical expression of it.

This field of research, which involves measurements and mathematics and the handling of large bodies of statistics, has been considerably cultivated, so that there is in existence in England a journal devoted exclusively to biometrics, which is edited by Carl Pearson, and is entitled *Biometrika*.

The whole subject of heredity is undergoing a thorough revision. What seems to be most needed at the present time is more exact experimentation, carried through several generations, together with more searching investigations into the microscopical constitution of egg and sperm, and close analysis of just what takes place during fertilization and the early stages of the development of the individual. Experiments are being conducted on an extended scale in endowed institutions. There is notably in this country, established under the Carnegie Institution, a station for experimental evolution, at Cold Spring Harbor, New York, of which C. B. Davenport is director. Other experimental stations in England and on the Continent have been established, and we are to expect as the result of coördinated and continuous experimental work many substantial contributions to the knowledge of inheritance.

CHAPTER XV

THE SCIENCE OF FOSSIL LIFE

It gradually dawned on the minds of men that the crust of the earth is like a gigantic mausoleum, containing within it the remains of numerous and varied forms of life that formerly existed upon the surface of the earth. The evidence is clear that untold generations of living forms, now preserved as fossils, inhabited the earth, disported themselves, and passed away long before the advent of man. The knowledge of this fossil life, on account of its great diversity, is an essential part of biology, and all the more so from the circumstance that many forms of life, remains of which are exhibited in the rocks, have long since become extinct. No history of biology would be complete without an account of the rise and progress of that department of biology which deals with fossil life.

It has been determined by collecting and systematically studying the remains of this ancient life that they bear testimony to a long, unbroken history in which the forms of both animals and plants have been greatly altered. The more ancient remains are simple in structure, and form with the later ones, a series that exhibits a gradually increasing complexity of structure. The study of the fossil series has brought about a very great extension of our knowledge regarding the age of the world and of the conditions under which life was evolved.

Strange Views Regarding Fossils.—But this state of our knowledge was a long time coming, and in the development of the subject we can recognize several distinct epochs, "well-marked by prominent features, but like all stages of intellectual growth, without definite boundaries." Fossils were known to the ancients, and by some of the foremost philosophers of Greece were understood to be the remains of animals and plants. After the revival of learning, however, lively controversies arose as to their nature and their meaning.

Some of the fantastic ideas that were entertained regarding the nature of fossil remains may be indicated. The fossils were declared by many to be freaks of nature; others maintained that they were the results of spontaneous generation, and were produced by the plastic forces of nature within the rocks in which they were found embedded. Another opinion expressed was that they were generated by fermentations. As the history of intellectual development shows, the mind has ever seemed benumbed in the face of phenomena that are completely misconceived; mystical explanations have accordingly been devised to account for them. Some of the pious persons of that period declared that fossils had been made and distributed by the Creator in pursuance of a plan beyond our comprehension. Another droll opinion expressed was that the Creator in His wisdom had introduced fossil forms into the rocks in order that they should be a source of confusion to the race of geologists that was later to arise.

And still another fantastic conception suggested that the fossils were the original molds used by the Creator in forming different varieties of animals and plants, some of which had been used and others discarded. It was supposed that in preparing for the creation of life He experimented and discarded some of His earliest attempts; and that fossils represented these discarded molds and also, perhaps, some that had been used in fashioning the created forms.

When large bones, as of fossil elephants, began to be exhumed, they became for the most part the objects of stupid wonder. The passage in the Scriptures was pointed out, that "there were giants in those days," and the bones were taken to be evidences of the former existence of giants. The opinions expressed regarding the fossil bones were varied and fantastic, "some saying that they were rained from Heaven, others saying that they were the gigantic limbs of the ancient patriarchs, men who were believed to be tall because they were known to be old." Following out this idea, "Henrion in 1718 published a work in which he assigned to Adam a height of 123 feet 9 inches, Noah being 20 feet shorter, and so on."

Determination of the Nature of Fossils.—In due course it came to be recognized that fossils were the remains of forms that had been alive during earlier periods of time; but in reaching this position there was continual controversy. Objections were especially vigorous from theological quarters, since such a conclusion was deemed to be contradictory to the Scriptures. The true nature of fossils had been clearly perceived by Leonardo da Vinci (1452–1519) and certain others in the sixteenth century.

The work, however, that approached more nearly to scientific demonstration was that of Steno (1638–1686), a Dane who migrated to Italy and became the court physician to the dukes of Tuscany. He was a versatile man who had laid fast hold upon the new learning of his day. Eminent as anatomist, physiologist, and physician, with his ever active mind he undertook to encompass all learning. It is interesting that Steno—or Stensen—after being passionately devoted to science, became equally devoted to religion and theology, and, forsaking all scientific pursuits, took orders and returned to his native country with the title of bishop. Here he worked in the service of humanity and religion to the end of his life.

In reference to his work in geology, his conclusions

regarding fossils (1669) were based on the dissection of the head of a shark, by which means he showed an almost exact correspondence between certain glossy fossils and the teeth of living sharks. He applied his reasoning, that like effects imply like causes, to all manner of fossils, and clearly established the point that they should be regarded as the remains of animals and plants. The method of investigation praeticed by Steno was that "which has consciously or unconsciously guided the researches of palæontologists ever since.'

Although his conclusions were well supported, they did not completely overthrow the opposing views, and become a fixed basis in geology. When, at the close of the eighteenth century and the beginning of the nineteenth, fossil remains were being exhumed in great quantities in the Paris basin, Cuvier, the great French naturalist, reëstablished the doctrine that fossils are the remains of ancient life. An account of this will be given presently, and in the mean time we shall go on with the consideration of a question raised by the conclusions of Steno.

Fossil Deposits Ascribed to the Flood.—After it began to be reluctantly conceded that fossils might possibly be the remains of former generations of animals and plants, there followed a period characterized by the general belief that these entombed forms had been deposited at the time of the Mosaic deluge. This was the prevailing view in the eighteenth century. As observation increased and the extent and variety of fossil life became known, as well as the positions in which fossils were found, it became more difficult to hold this view with any appearance of reason. Large forms were found on the tops of mountains, and also lighter forms were found near the bottom. Miles upon miles of superimposed rocks were discovered, all of them bearing quantities of animal forms, and the interpretation that these had been killed and distributed by a deluge became very strained. But to the reasoners who gave free play to their fancies the facts of observation afforded little difficulty. Some declared that the entire surface of the earth had been reduced to the condition of a pasty mass, and that the animals drowned by the Deluge had been deposited within this pasty mass which, on the receding of the waters, hardened into rocks.

The belief that fossil deposits were due to the Deluge sensibly declined, however, near the close of the eighteenth century, but was still warmly debated in the early part of the nineteenth century. Fossil bones of large tropical animals having been discovered about 1821, embedded in the stalagmite-covered floor of a cavern in Yorkshire, England, some of the ingenious supporters of the flood-theory maintained that caves were produced by gases proceeding from the bodies of decaying animals of large size; that they were like large bubbles in the crust of the earth, and, furthermore, that bones found in caverns were either those from the decayed carcasses or others that had been deposited during the occurrence of the Flood.

Even the utterances of Cuvier, in his theory of catastrophism to which we shall presently return, gave countenance to the conclusion that the Deluge was of universal extent. As late as 1823, William Buckland, reader in geology in Oxford, and later canon (1825) of Christ Church, and dean (1845) of Westminster, published his *Reliquiæ Diluvianæ*, or *Observations on the Organic Remains Attesting the Action of a Universal Deluge.*

The theory that the Mosaic deluge had any part in the deposit of organic fossils was finally surrendered through the advance of knowledge, owing mainly to the labors of Lyell and his followers.

The Comparison of Fossil and Living Animals.—The very great interest connected with the reëstablishment of the conclusion of Steno, that fossils were once alive, leads us to speak more at length of the discoveries upon which Cuvier passed his opinion. In the gypsum rocks about Paris the workmen had been turning up to the light bones of enormous size. While the workmen could recognize that they were bones of some monsters, they were entirely at loss to imagine to what kind of animals they had belonged, but the opinion was frequently expressed that they were the bones of human giants.

Cuvier, with his extensive preparation in comparative anatomy, was the best fitted man perhaps in all the world to pass judgment upon these particular bones. He went to the quarries and, after observing the remains, he saw very clearly that they were different from the bones of any animals now existing. His great knowledge of comparative anatomy was founded on a comprehensive study of the bony system as well as the other structures of all classes of living animals. He was familiar with the anatomy of elephants, and when he examined the large bones brought to light in the quarries of Montmartre, he saw that he was confronted with the bones of elephant-like animals, but animals differing in their anatomy from those at present living on the earth.

The great feature of Cuvier's investigations was that he instituted comparisons on a broad scale between fossil remains and living animals. It was not merely that he followed the method of investigation employed by Steno; he went much further and reached a new conclusion of great importance. Not only was the nature of fossil remains determined, but by comparing their structure with that of living animals the astounding inference was drawn that the fossil remains examined belonged to forms that were truly extinct. This discovery marks an epoch in the development of the knowledge of extinct animals.

Cuvier the Founder of Vertebrate Palæontology.—The interesting discovery that the fossil relics in the Eocene rocks about Paris embraced extinct species was announced to the Institute by Cuvier in January, 1796; and thereafter he continued for a quarter of a century to devote much attention to the systematic study of collections made in that district. These observations were, however, shared with other labors upon comparative anatomy and zoölogy, which indicates the prodigious industry for which he was notable. In 1812–1813 he published a monumental work, profusely illustrated, under the title *Ossemens Fossiles.* This standard publication entitles him to recognition as the founder of vertebrate palæontology.

In examining the records of fossil life, Cuvier and others saw that the evidence indicated a succession of animal populations that had become extinct, and also that myriads of new forms of life appeared in the rocks of succeeding ages. Here Cuvier, who believed that species were fixed and unalterable, was confronted with a puzzling problem. In attempting to account for the extinction of life, and what seemed to him the creation of new forms, he could see no way out consistent with his theoretical views except to assume that the earth had periodically been the scene of great catastrophes, of which the Mosaic deluge was the most recent, but possibly not the last. He supposed that these cataclysms of nature resulted in the extinction of all life, and that after each catastrophe the salubrious condition of the earth was restored, and that it was re-peopled by a new creation of living beings. This conception, known as the theory of catastrophism, was an obstacle to the progress of science. It is to be regretted that Cuvier was not able to accept the views of his illustrious contemporary Lamarck, who believed that the variations in fossil life, as well as those of living forms, were owing to gradual transformations.

Lamarck Founds Invertebrate Palæontology.—The credit of founding the science of palæontology does not belong exclusively to Cuvier. Associated with his name as cofounders are those of Lamarck and William Smith. Lamarck, that quiet, forceful thinker who for so many years worked by the side of Cuvier, founded the science of invertebrate palæontology. The large bones with which Cuvier worked were more easy to be recognized as unique or as belonging to extinct animals than the shells which occurred in abundance in the rocks about Paris. The latter were more difficult to place in their true position because the number of forms of life in the sea is very extended and very diverse. Just as Cuvier was a complete master of knowledge regarding vertebrate organization, so Lamarck was equally a master of that vast domain of animal forms which are of a lower grade of organization—the invertebrates. From his study of the collections of shells and other invertebrate forms from the rocks, Lamarck created invertebrate palæontology and this, coupled with the work of Cuvier, formed the foundations of the entire field.

Lamarck's study of the extinct invertebrates led him to conclusions widely at variance with those of Cuvier. Instead of thinking of a series of catastrophes, he saw that not all of the forms of life belonging to one geological period became extinct, but that some of them were continued into the succeeding period. He saw, therefore, that the succession of life in the rocks bore testimony to a long series of gradual changes upon the earth's surface, and did not in any way indicate the occurrence of catastrophes. The changes, according to the views of Lamarck, were all knit together into a continuous process, and his conception of the origin of life upon the earth grew and expanded until it culminated in the elaboration of the first consistent theory of evolution.

These two men, Lamarck and Cuvier, form a contrast as to the favors distributed by fortune: Cuvier, picturesque, highly honored, the favorite of princes, advanced to the highest places of recognition in the government, acclaimed as the Jove of natural science; Lamarck, hard-working, harassed by poverty, insufficiently recognized, and, although more gifted than his confrère, overlooked by the scientific men of the time. The judgment of the relative position of these two men in natural science is now being reversed, and on the basis of intellectual supremacy Lamarck is coming into general recognition as the better man of the two. In the chapters dealing with organic evolution some events in the life of this remarkable man will be given.

The Arrangement of Fossils in Strata.—The other name associated with Lamarck and Cuvier is that of William Smith, the English surveyor. Both Lamarck and Cuvier were men of extended scientific training, but William Smith had a moderate education as a surveyor. While the two former were able to express scientific opinions upon the nature of the fossil forms discovered, William Smith went at his task as an observer with a clear and unprejudiced mind, an observer who walked about over the fields, noticing the conditions of rocks and of fossil forms embedded therein. He noted that the organic remains were distributed in strata, and that particular forms of fossil life characterized particular strata and occupied the same relative position to one another. He found, for illustration, that certain particular forms would be found underlying certain other forms in one mass of rocks in a certain part of the country. Wherever he traveled, and whatever rocks he examined, he found these forms occupying the same relative positions, and thus he came to the conclusion that the living forms within the rocks constitute a stratified series, having definite and unvarying arrangement with reference to one another.

In short, the work of these three men—Cuvier, Lamarck, and William Smith—placed the new science of palæontology upon a secure basis at the beginning of the nineteenth century.

Summary.—The chief steps up to this time in the growth of the science of fossil life may now be set forth in categories, though we must remember that the advances proceeded concurrently and were much intermingled, so that, whatever arrangement we may adopt, it does not represent a strict chronological order of events:

I. The determination of the nature of fossils. Owing to the labors of Da Vinci, Steno, and Cuvier, the truth was established that fossils are the remains of former generations of animals and plants.

II. The comparison of organic fossils with living forms that was instituted on a broad scale by Cuvier resulted in the conclusion that some of the fossils belong to extinct races. The belief of Cuvier that entire populations became extinct simultaneously, led him to the theory of catastrophism. The observations of Lamarck, that, while some species disappear, others are continued and pass through transmutations, were contrary to that theory.

III. The recognition that the stratified rocks in which fossils are distributed are sedimentary deposits of gradual formation. This observation and the following took the ground from under the theory that fossils had been deposited during the Mosaic deluge.

IV. The discovery by William Smith that the arrangement of fossils within rocks is always the same, and the relative age of rocks may be determined by an examination of their fossil contents.

Upon the basis of the foregoing, we come to the next advance, *viz.:*

V. The application of this knowledge to the determination of the history of the earth.

Fossil Remains as an Index to the Past History of the Earth.—The most advanced and enlightened position that had been taken in reference to the fossil series during the first third of the nineteenth century was that taken by Lamarck, he being the first to read in the series the history of life upon the globe, weaving it into a connected story, and establishing thereon a doctrine of organic evolution. It was not until after 1859, however, that the truth of this conclusion was generally admitted, and when it was accepted it was not through the earlier publications of Lamarck, but through the arguments of later observers, founded primarily upon the hypothesis set forth by Darwin. There were several gradations of scientific opinion in the period, short as it was, between the time of Cuvier and of Darwin; and this intermediate period was one of contention and warfare between the theologians and the geologists. Cuvier had championed the theory of a succession of catastrophes, and since this hypothesis did not come into such marked conflict with the prevailing theological opinion as did the views of Lamarck, the theologians were ready to accept the notion of Cuvier, and to point with considerable satisfaction to his unique position as an authority.

Lyell.—In 1830 there was published an epoch-making work in geology by Charles Lyell (Fig. 97), afterward Sir Charles, one of the most brilliant geologists of all the world. This British leader of scientific thought showed the prevalence of a uniform law of development in reference to the earth's surface. He pointed out the fact that had been maintained by Hutton, that changes in the past were to be interpreted in the light of what is occurrıng in the present. By making a careful study of the work performed by the waters in cutting down the continents and in transferring the eroded material to other places, and distributing it in the form of deltas; by observing also the action of frost and wind and wave; by noting, furthermore, the conditions under which animals die and are subsequently covered up in the matrix of detritus—by all this he showed evidences of a series of slow, continuous changes that have occurred in the past and have molded the earth's crust into its present condition.

He showed, further, that organic fossils are no exception to this law of uniform change. He pointed to the evidences that ages of time had been required for the formation of the rocks bearing fossils; and that the regular succession of animal

FIG. 97.—CHARLES LYELL, 1797–1875.

forms indicates a continual process of development of animal life; and that the disappearance of some forms, that is, their becoming extinct, was not owing to sudden changes, but to gradual changes. When this view was accepted, it overthrew the theory of catastrophism and replaced it by one designated uniformatism, based on the prevalence of uniform natural laws.

This new conception, with all of its logical inferences,

was scouted by those of theological bias, but it won its way in the scientific world and became an important feature in preparing for the reception of Darwin's great book upon the descent of animal life.

We step forward now to the year 1859, to consider the effect upon the science of palæontology of the publication of Darwin's *Origin of Species*. Its influence was tremendous. The geological theories that had provoked so much controversy were concerned not merely with the disappearance of organic forms, but also with the introduction of new species. The *Origin of Species* made it clear that the only rational point of view in reference to fossil life was that it had been gradually developed, that it gave us a picture of the conditions of life upon the globe in past ages, that the succession of forms within the rocks represented in outline the successive steps in the formation of different kinds of animals and plants.

Owen.—Both before and after Darwin's hypothesis was given to science, notable anatomists, a few of whom must be mentioned, gave attention to fossil remains. Richard Owen (1804–1892) had his interest in fossil life stimulated by a visit to Cuvier in 1831, and for more than forty years thereafter he published studies on the structure of fossil animals. His studies on the fossil remains of Australia and New Zealand brought to light some interesting forms. The extinct giant bird of New Zealand (Fig. 98) was a spectacular demonstration of the enormous size to which birds had attained during the Eocene period. Owen's monograph (1879) on the oldest known bird—the archæopteryx—described an interesting form uniting both bird-like and reptilian characteristics.

Agassiz.—Louis Agassiz (1807–1873) (Fig. 99) also came into close personal contact with Cuvier, and produced his first great work partly under the stimulus of the latter. When

FIG. 98.—Professor Owen and the Extinct Fossil Bird (Dinornis) of New Zealand.

Permission of D. Appleton & Co.

Agassiz visited Paris, Cuvier placed his collections at Agassiz's disposal, together with numerous drawings of fossil fishes. The profusely illustrated monograph of Agassiz on the fossil fishes (1833–1844) began to appear in 1833, the year after

FIG. 99.—LOUIS AGASSIZ, 1807–1873.

Cuvier's death, and was carried on eleven years before it was completed.

Agassiz, with his extensive knowledge of the developmental stages of animals, came to see a marked parallelism between the stages in development of the embryo and the successive forms in the geological series. This remarkable parallelism between the fossil forms of life and the stages in the development of higher forms of recent animals is very interesting and very significant, and helps materially in elucidating the idea that the fossil series represent roughly the successive stages through which animal forms have passed in their upward course of development from the simplest to the highest, through long ages of time. Curiously enough, however, Agassiz failed to grasp the meaning of the principle that he had worked out. After illustrating so nicely the process of organic evolution, he remained to the end of his life an opponent of that theory.

Huxley.—Thomas Henry Huxley (1825–1895) was led to study fossil life on an extended scale, and he shed light in this province as in others upon which he touched. With critical analysis and impartial mind he applied the principle of evolution to the study of fossil remains. His first conclusion was that the evidence of evolution derived from palæontology was negative, but with the advances in discovery he grew gradually to recognize that palæontologists, in bringing to light complete evolutionary series, had supplied some of the strongest supporting evidence of organic evolution. By many geologists fossils have been used as time-markers for the determination of the age of various deposits; but, with Huxley, the study of them was always biological. It is to the latter point of view that palæontology owes its great importance and its great development. The statement of Huxley, that the only difference between a fossil and a recent animal is that one has been dead longer than the other, represents the spirit in which the study is being carried forward.

With the establishment of the doctrine of organic evolution palæontology entered upon its modern phase of growth· upon this basis there is being reared a worthy structure through the efforts of the recent votaries to the science. It is neither essential nor desirable that the present history of the subject should be followed here in detail. The collections of material upon which palæontologists are working have been enormously increased, and there is perhaps no place where activity has been greater than in the United States. The rocks of the Western States and Territories

FIG. 100.—E. D. COPE, 1840–1897.

embrace a very rich collection of fossil forms, and, through the generosity of several wealthy men, exploring parties have been provided for and immense collections have been brought back to be preserved in the museums, especially of New Haven, Conn., and in the American Museum of Natural History in New York City.

Leidy, Cope, and Marsh.—Among the early explorers of the fossils of the West must be named Joseph Leidy, E. D. Cope (Fig. 100), and O. C. Marsh. These gentlemen all had access to rich material, and all of them made notable contributions to the science of palæontology. The work of

FIG. 101.—O. C. MARSH, 1831–1899.

Cope (1840–1897) is very noteworthy. He was a comparative anatomist equal to Cuvier in the extent of his knowledge, and of larger philosophical views. His extended publications under the direction of the United States Government have very greatly extended the knowledge of fossil vertebrate life in America.

O. C. Marsh (Fig. 101) is noteworthy for similar explorations; his discovery of toothed birds in the Western rocks and his collection of fossil horses, until recently the most complete one in existence, are all very well known. Throughout his long life he contributed from his own private fortune, and intellectually through his indefatigable labors, to the progress of palæontology.

Zittel.—The name most widely known in palæontology is that of the late Karl von Zittel (1839–1904), who devoted all his working life to the advancement of the science of fossils. In his great work, *Handbuch der Palaeontologie* (1876–1893), he brought under one view the entire range of fossils from the protozoa up to the mammals. Osborn says: "It is probably not an exaggeration to say that he did more for the promotion and diffusion of palæontology than any other single man who lived during the nineteenth century. While not gifted with genius, he possessed extraordinary judgment, critical capacity, and untiring industry." His portrait (Fig. 102) shows a face "full of keen intelligence and enthusiasm."

Zittel's influence was exerted not only through his writings, but also through his lectures and the stimulus imparted to the large number of young men who were attracted to Munich to study under his direction. These disciples are now distributed in various universities in Europe and the United States, and are there carrying forward the work begun by Zittel. The great collection of fossils which he left at Munich contains illustrations of the whole story of the evolution of life through geological ages.

Recent Developments.—The greatest advance now being made in the study of fossil vertebrate life consists in establishing the lineage of families, orders, and classes. Investigators have been especially fortunate in working out the direct line of descent of a number of living mammals. Fossils have been collected which supply a panoramic view of the line of descent of horses, of camels, of rhinoceroses, and of other animals. The most fruitful worker in this field at the present time is perhaps Henry F. Osborn, of the American Museum

FIG. 102.—KARL VON ZITTEL, 1839–1904.

of Natural History, New York City. His profound and important investigations in the ancestry of animal life are now nearing the time of their publication in elaborated form.

Palæontology, by treating fossil life and recent life in the same category, has come to be one of the important lines of investigation in biology. It is, of course, especially rich in giving us a knowledge of the hard parts of animals, but by ingenious methods we can arrive at an idea of some of the soft parts that have completely disappeared. Molds of the interior of the cranium can be made, and thus one may form a notion of the relative size and development of the brain in different vertebrated animals. This method of making molds and studying them has shown that one characteristic of the geological time of the tertiary period was a marked development in regard to the brain size of the different animals. There was apparently, just prior to the quaternary epoch, a need on the part of animals to have an increased brain-growth; and one can not doubt that this feature which is demonstrated by fossil life had a great influence in the development of higher animal forms.

The methods of collecting fossils in the field have been greatly developed. By means of spreading mucilage and tissue paper over delicate bones that crumble on exposure to the air, and of wrapping fossils in plaster casts for transportation, it has been made possible to uncover and preserve many structures which with a rougher method of handling would have been lost to science.

Fossil Man.—One extremely interesting section of palæontology deals with the fossil remains of the supposed ancestors of the present human race. Geological evidence establishes the great antiquity of man, but up to the present time little systematic exploration has been carried on with a view to discover all possible traces of fossil man. From time to time since 1840 there have been discovered in caverns and river-gravels bones which, taken together, constitute an interesting series. The parts of the skull are of especial importance in this kind of study, and there now exists in different collections a series containing the Neanderthal skull, the skulls of Spy and Engis, and the Java skull described in 1894 by Dubois. There have also been found recently (November, 1906) in deposits near Lincoln, Neb., some fossil human remains that occupy an intermediate position between the Neanderthal skull and the skulls of the lower representatives of living races of mankind. We shall have occasion to revert to this question in considering the evidences of organic evolution. (See page 364.)

The name palæontology was brought into use about 1830. The science affords, in some particulars, the most interesting field for biological research, and the feature of the reconstruction of ancient life and the determination of the lineage of living forms has taken a strong hold on the popular imagination. According to Osborn, the most important palæontological event of recent times was the discovery, in 1900, of fossil beds of mammals in the Fayûm lake-province of Egypt, about forty-seven miles south of Cairo. Here are embedded fossil forms, some of which have been already described in a volume by Charles W. Andrews, which Osborn says "marks a turning-point in the history of mammalia of the world." It is now established that "Africa was a very important center in the evolution of mammalian life." It is expected that the lineage of several orders of mammalia will be cleared up through the further study of fossils from this district.

PART II

THE DOCTRINE OF ORGANIC EVOLUTION

CHAPTER XVI

WHAT EVOLUTION IS: THE EVIDENCE UPON WHICH IT RESTS, ETC.

The preceding pages have been devoted mainly to an account of the shaping of ideas in reference to the architecture, the physiology, and the development of animal life.

We come now to consider a central theme into which all these ideas have been merged in a unified system; *viz.*, the process by which the diverse forms of animals and plants have been produced.

Crude speculations regarding the derivation of living forms are very ancient, and we may say that the doctrine of organic evolution was foreshadowed in Greek thought. The serious discussion of the question, however, was reserved for the nineteenth century. The earlier naturalists accepted animated nature as they found it, and for a long time were engaged in becoming acquainted merely, with the different kinds of animals and plants, in working out their anatomy and development; but after some progress had been made in this direction there came swinging into their horizon deeper questions, such as that of the derivation of living forms. The idea that the higher forms of life are derived from simpler ones by a process of gradual evolution received general acceptance, as we have said before, only in the last part of the nineteenth century, after the work of Charles Darwin; but we shall presently see how the theory of organic development was thought out in completeness by

Lamarck in the last years of the eighteenth century, and was further molded by others before Darwin touched it.

Vagueness Regarding Evolution.—Although "evolution" is to-day a word in constant use, there is still great vagueness in the minds of most people as to what it stands for; and, what is more, there is very little general information disseminated regarding the evidence by which it is supported, and regarding the present status of the doctrine in the scientific world.

In its broad sense, evolution has come to mean the development of all nature from the past. We may, if we wish, think of the long train of events in the formation of the world, and in supplying it with life as a story inscribed upon a scroll that is being gradually unrolled. Everything which has come to pass is on that part so far exposed, and everything in the future is still covered, but will appear in due course of time; thus the designation of evolution as "the unrolling of the scroll of the universe" becomes picturesquely suggestive. In its wide meaning, it includes the formation of the stars, solar systems, the elements of the inorganic world, as well as all living nature—this is general evolution; but the word as commonly employed is limited to organic evolution, or the formation of life upon our planet. It will be used hereafter in this restricted sense.

The vagueness regarding the theory of organic evolution arises chiefly from not understanding the points at issue. One of the commonest mistakes is to confuse Darwinism with organic evolution. It is known, for illustration, that controversies are current among scientific workers regarding Darwinism and certain phases of evolution, and from this circumstance it is assumed that the doctrine of organic evolution as a whole is losing ground. The discussions of De Vries and others—all believers in organic evolution—at the Scientific Congress in St. Louis in 1904, led to the statement in the public press that the scientific world was haggling over the evolution-theory, and that it was beginning to surrender it. Such statements are misleading and tend to perpetuate the confusion regarding its present status. Furthermore, the matter as set forth in writings like the grotesque little book, *At the Deathbed of Darwinism* tends to becloud rather than to clear the atmosphere.

The theory of organic evolution relates to the history of animal and plant life, while Darwin's theory of natural selection is only one of the various attempts to point out the causes for that history's being what it is. An attack upon Darwinism is not, in itself, an attack upon the general theory, but upon the adequacy of his explanation of the way in which nature has brought about the diversity of animal and plant life. Natural selection is the particular factor which Darwin has emphasized, and the discussion of the part played by other factors tends only to extend the knowledge of the evolutionary process, without detracting from it as a general theory.

While the controversies among scientific men relate for the most part to the influences that have been operative in bringing about organic evolution, nevertheless there are a few in the scientific camp who repudiate the doctrine. Fleischmann, of Erlangen, is perhaps the most conspicuous of those who are directing criticism against the general doctrine, maintaining that it is untenable. Working biologists will be the first to admit that it is not demonstrated by indubitable evidence, but the weight of evidence is so compelling that scientific men as a body regard the doctrine of organic evolution as merely expressing a fact of nature, and we can not in truth speak of any considerable opposition to it. Since Fleischmann speaks as an anatomist, his suppression of anatomical facts with which he is acquainted and his form of special pleading have impressed the biological world as lacking in sincerity.

This is not the place, however, to deal with the technical aspects of the discussion of the factors of organic evolution; it is rather our purpose here to give a descriptive account of the theory and its various explanations. First we should aim to arrive at a clear idea of what the doctrine of evolution is, and the basis upon which it rests; then of the factors which have been emphasized in attempted explanations of it; and, finally, of the rise of evolutionary thought, especially in the nineteenth century. The bringing forward of these points will be the aim of the following pages.

Nature of the Question.—It is essential at the outset to perceive the nature of the question involved in the theories of organic evolution. It is not a metaphysical question, capable of solution by reflection and reasoning with symbols; the data for it must rest upon observation of what has taken place in the past in so far as the records are accessible. It is not a theological question, as so many have been disposed to argue, depending upon theological methods of interpretation. It is not a question of creation through divine agencies, or of non-creation, but a question of method of creation.

Evolution as used in biology is merely a history of the steps by which animals and plants came to be what they are. It is, therefore, a historical question, and must be investigated by historical methods. Fragments of the story of creation are found in the strata of the earth's crust and in the stages of embryonic development. These clues must be brought together; and the reconstruction of the story is mainly a matter of getting at the records. Drummond savs that evolution is "the story of creation as told by those who know it best."

The Historical Method.—The historical method as applied to searching out the early history of mankind finds a parallel in the investigations into the question of organic evolution. In the buried cities of Palestine explorers have uncovered traces of ancient races and have in a measure reconstructed their history from fragments, such as coins, various objects of art and of household use, together with inscriptions on tombs and columns and on those curious little bricks which were used for public records and correspondence. One citv having been uncovered, it is found by lifting the floors of temples and other buildings, and the pavement of public squares, that this city, although very ancient, is built upon the ruins of a more ancient one, which in turn covers the ruins of one still older. In this way, as many as seven successive cities have been found, built one on top of the other, and new and unexpected facts regarding ancient civilization have been brought to light. We must admit that this gives us an imperfect history, with many gaps; but it is one that commands our confidence, as being based on facts of observation, and not on speculation.

In like manner the knowledge of the past history of animal life is the result of explorations by trained scholars into the records of the past. We have remains of ancient life in the rocks, and also traces of past conditions in the developing stages of animals. These are all more ancient than the inscriptions left by the hand of man upon his tombs, his temples, and his columns, but nevertheless full of meaning if we can only understand them. This historical method of investigation applied to the organic world has brought new and unexpected views regarding the antiquity of life.

The Diversity of Living Forms.—Sooner or later the question of the derivation of the animals and plants is bound to come to the mind of the observer of nature. There exist at present more than a million different kinds of animals. The waters, the earth, the air teem with life. The fishes of the sea are almost innumerable, and in a single order of the insect-world, the beetles, more than 50,000 species are known and described. In addition to living animals, there is entombed in the rocks a great multitude of fossil forms which lived centuries ago, and many of which have become entirely extinct. How shall this great diversity of life be accounted for? Has the great variety of forms existed unchanged from the days of their creation to the present? Or have they, perchance, undergone modifications so that one original form, or at least a few original types, may have through transformations merged into different kinds? This is not merely an idle question, insoluble from the very nature of the case; for the present races of animals have a lineage reaching far into the past, and the question of fixity of form as against alteration of type is a historical question, to be answered by getting evidence as to their line of descent.

Are Species Fixed in Nature?—The aspect of the matter which presses first upon our attention is this: Are the species (or different kinds of animals and plants) fixed, and, within narrow limits, permanent, as Linnæus supposed? Have they preserved their identity through all time, or have they undergone changes? This is the heart of the question of organic evolution. If observation shows species to be constant at the present time, and also to have been continuous so far as we can trace their parentage, we must conclude that they have not been formed by evolution; but if we find evidence of their transmutation into other species, then there has been evolution.

It is well established that there are wide ranges of variation among animals and plants, both in a wild state and under domestication. Great changes in flowers and vegetables are brought about through cultivation, while breeders produce different kinds of pigeons, fowls, and stock. We know, therefore, that living beings may change through modification of the circumstances and conditions that affect their lives. But general observations extending over a few decades are not sufficient. We must, if possible, bring the history of past ages to bear upon the matter, and determine whether or not there had been, with the lapse of time, any considerable alteration in living forms.

Evolutionary Series.—Fortunately, there are preserved in the rocks the petrified remains of animals, showing their history for many thousands of years, and we may use them to test the question. It is plain that rocks of a lower level were deposited before those that cover them, and we may safely assume that the fossils have been preserved in their proper chronological order. Now, we have in Slavonia some fresh-water lakes that have been drying up from the tertiary period. Throughout the ages, these waters were inhabited by snails, and naturally the more ancient ones were the parents of the later broods. As the animals died their shells sank to the bottom and were covered by mud and débris, and held there like currants in a pudding. In the course of ages, by successive accumulations, these layers thickened and were changed into rock, and by this means shells have been preserved in their proper order of birth and life, the most ancient at the bottom and the newest at the top. We can sink a shaft or dig a trench, and collect the shells and arrange them in proper order.

Although the shells in the upper strata are descended from those near the bottom, they are very different in appearance. No one would hesitate to name them different species; in fact, when collections were first made, naturalists classified these shells into six or eight different species. If, however, a collection embracing shells from all levels is arranged in a long row in proper order, a different light is thrown on the matter; while those at the ends are unlike, yet if we begin at one end and pass to the other we observe that the shells all grade into one another by such slight changes that there is no line showing where one kind leaves off and another begins. Thus their history for thousands of years bears testimony to the fact that the species have not remained constant, but have changed into other species.

Fig. 103 will give an idea of the varieties and gradations. It represents shells of a genus, Paludina, which is still abundant in most of the fresh waters of our globe.

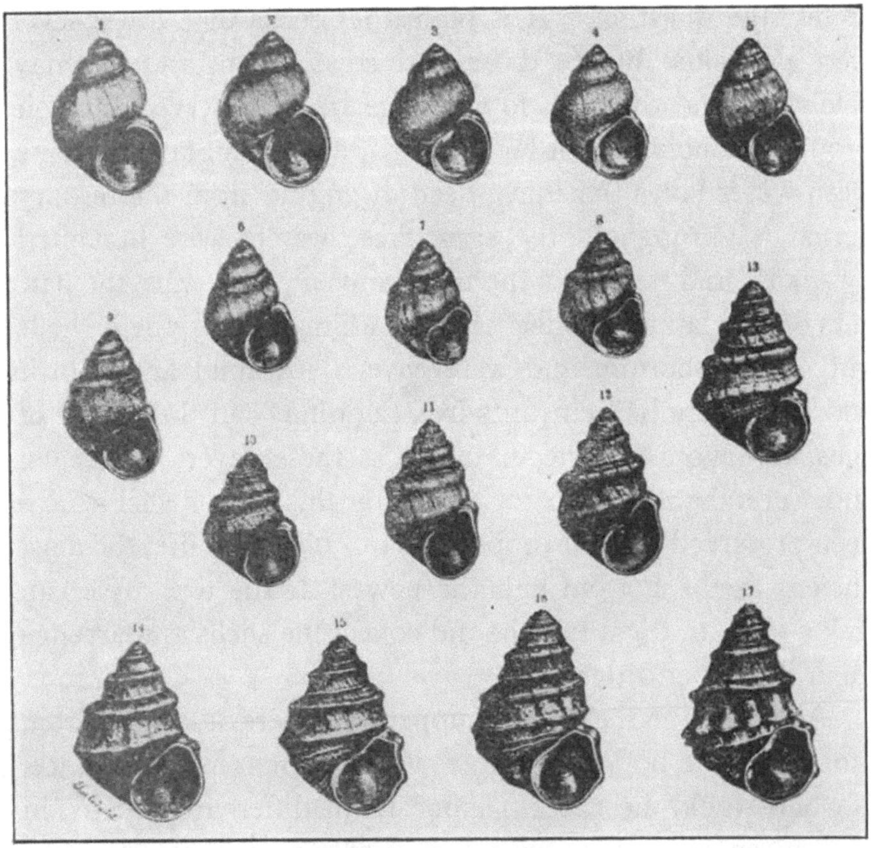

FIG. 103.—Transmutations of Paludina. (After Neumayer.)

A similar series of shells has been brought to light in Württemberg in which the variations pass through wider limits, so that not only different species may be observed, but different genera connected by almost insensible gradations. These transformations are found in a little flattened pond-shell similar to the planorbis, which is so common at the present time.

Fig. 104 shows some of these transformations, the finer

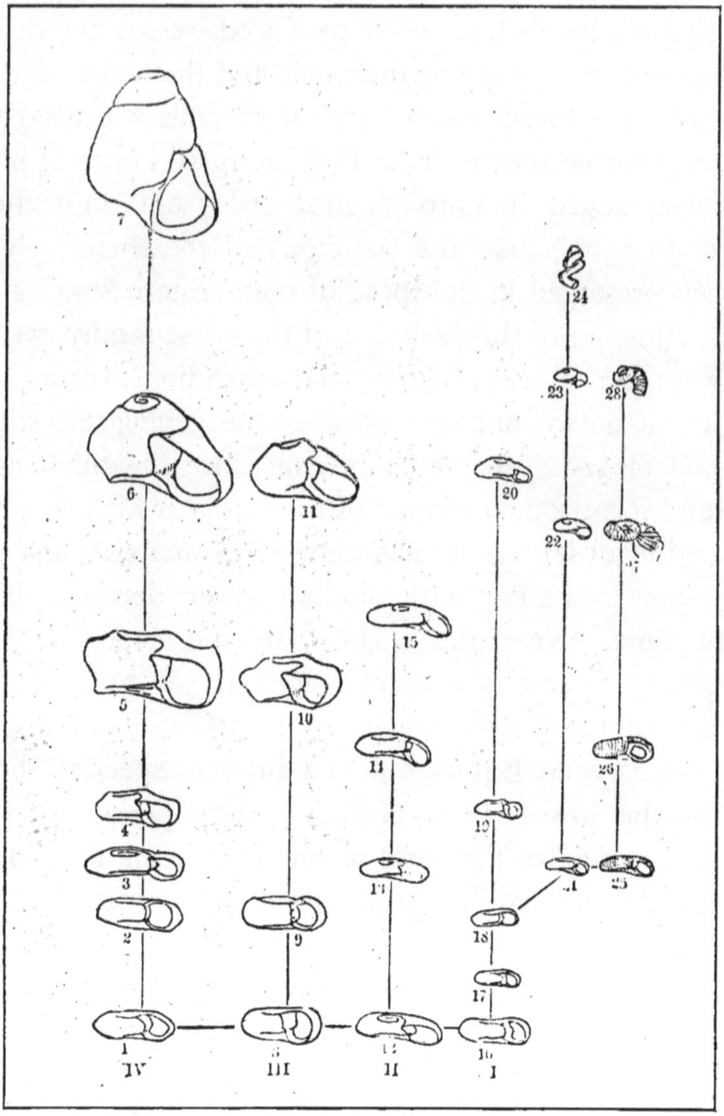

FIG. 104.—Planorbis Shells from Steinheim. (After Hyatt.)

gradations being omitted. The shells from these two sources bear directly upon the question of whether or not species have held rigidly to their original form.

After this kind of revelation in reference to lower animals, we turn with awakened interest to the fossil bones of the higher animals.

Evolution of the Horse.—When we take into account the way in which fossils have been produced we see clearly that it is the hard parts, such as the shells and the bones, that will be preserved, while the soft parts of animals will disappear. Is it not possible that we mav find the fossil bones of higher animals arranged in chronological order and in sufficient number to supplement the testimony of the shells? There has been preserved in the rocks of our Western States a very complete history of the evolution of the horse family, written, as it were, on tablets of stone, and extending over a period of more than two million years, as the geologists estimate time. Geologists can, of course, measure the thickness of rocks and form some estimate of the rate at which they were deposited by observing the character of the material and comparing the formation with similar water deposits of the present time. Near the surface, in the deposits of the quarternary period, are found remains of the immediate ancestors of the horse, which are recognized as belonging to the same genus, Equus, but to a different species; thence, back to the lowest beds of the tertiary period we come upon the successive ancestral forms, embracing several distinct genera and exhibiting an interesting series of transformations.

If in this way we go into the past a half-million years, we find the ancestors of the horse reduced in size and with three toes each on the fore and hind feet. The living horse now has only a single toe on each foot, but it has small splint-like bones that represent the rudiments of two more. If we go back a million years, we find three toes and the rudiments of a fourth; and going back two million years, we find four fully developed toes, and bones in the feet to support them.

It is believed that in still older rocks a five-toed form will be discovered, which was the parent of the four-toed form.

In the collections at Yale College there are preserved upward of thirty steps or stages in the history of the horse family, showing that it arose bv evolution or gradual change from a four- or five-toed ancestor of about the size of a fox, and that it passed through many changes, besides increase in size, in the two million years in which we can get facts as to its history.

Remarkable as is this feature of the Marsh collection at New Haven, it is now surpassed by that in the Museum of Natural History in New York City. Here, through the munificent gifts of the late W. C. Whitney, there has been accumulated the most complete and extensive collection of fossil horses in the world. This embraced, in 1904, some portions of 710 fossil horses, 146 having been derived from explorations under the Whitney fund. The extraordinary character of the collection is shown from the fact that it contains five complete skeletons of fossil horses—more than existed at that time in all other museums of the world.

The specimens in this remarkable collection show phases in the parallel development of three or four distinct races of horse-like animals, and this opens a fine problem in comparative anatomy; *viz.*, to separate those in the direct line of ancestry of our modern horse from all the others. This has been accomplished by Osborn, and through his critical analysis we have become aware of the fact that the races of fossil horses had not been distinguished in any earlier studies. As a result of these studies, a new ancestry of the horse, differing in details from that given by Huxley and Marsh, is forthcoming.

Fig. 105 shows the bones of the foreleg of the modern horse, and Fig. 106 some of the modifications through which it has passed. Fig. 107 shows a reconstruction of the ancestor of the horse made by Charles R. Knight, the animal painter, under the direction of Professor Osborn.

While the limbs were undergoing the changes indicated, other parts of the organism were also being transformed

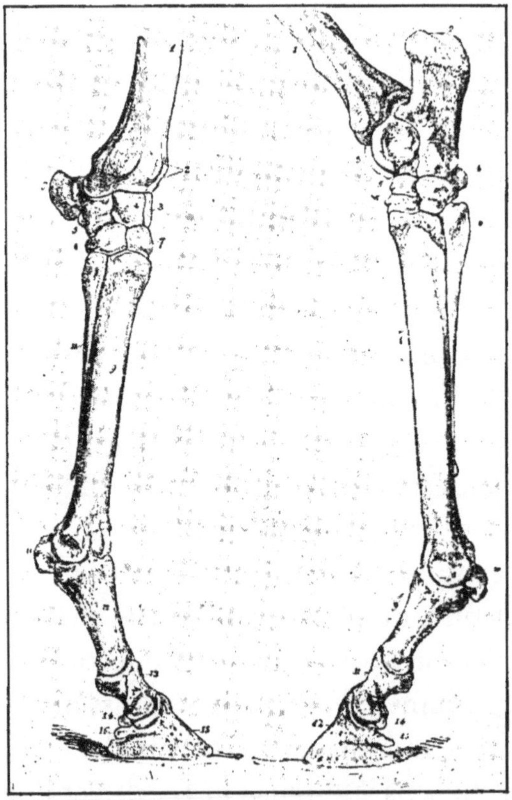

FIG. 105.—Bones of the Foreleg of a Horse.

and adapted to the changing conditions of its life. The evolution of the grinding teeth of the horse is fully exhibited in the fossil remains. All the facts bear testimony that the horse was not originally created as known to-day, but that his ancestors existed in different forms, and in evolution have transcended several genera and a considerable number of species. The highly specialized limb of the horse adapted for speed was the product of a long series of changes, of which the record is fairly well preserved. Moreover, the records show that the atavus of the horse began in North America, and that by migration the primitive horses spread from this continent to Europe, Asia, and Africa.

So far we have treated the question of fixity of species as a historical one, and have gone searching for clues of past

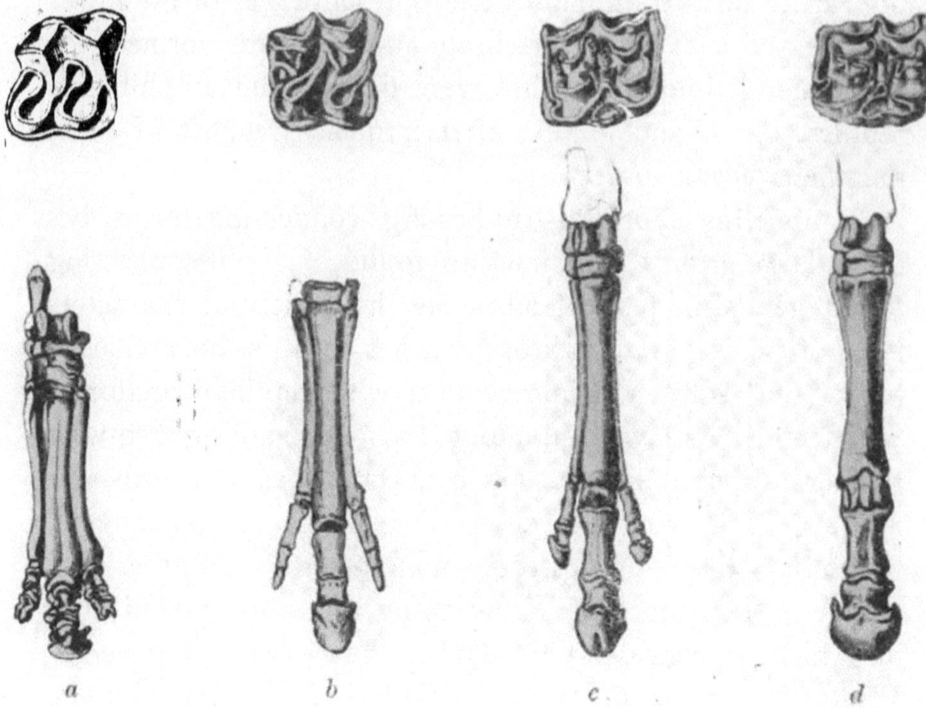

FIG. 106.—Bones of the Foreleg and Molar Teeth of Fossil Ancestors of the Horse. European Forms. (After Kayser.)

conditions just as an archæologist explores the past in buried cities. The facts we have encountered, taken in connection with a multitude of others pointing in the same direction, begin to answer the initial question, Were the immense numbers of living forms created just as we find them, or were they evolved by a process of transformation?

The geological record of other families of mammals has also been made out, but none so completely as that of the horse family. The records show that the camels were native in North America, and that they spread by migration from the land of their birth to Asia and Africa, probably crossing by means of land-connections which have long since become submerged.

The geological record, considered as a whole, shows that the earlier formed animals were representatives of the lower groups, and that when vertebrate animals were formed, for a very long time only fishes were living, then amphibians, reptiles, birds, and finally, after immense reaches of time, mammals began to appear.

Connecting Forms.—Interesting connecting forms between large groups sometimes are found, or, if not connecting forms, generalized ones embracing the structural characteristics of two separate groups. Such a form is the archæopteryx (Fig. 108), a primitive bird with reptilian anatomy, with teeth in its jaws, and a long, lizard-like tail covered with feathers, which seems to show connection between birds and reptiles. The wing also shows the supernumerary fingers, which have been suppressed in modern birds. Another suggestive type of this kind is the flying reptile or pterodactyl, of which a considerable number have been discovered. Illustrations indicating that animals have had a common line of descent might be greatly multiplied.

The Embryological Record and its Connection with Evolution.—The most interesting, as well as the most comprehensive clues bearing on the evolution of animal life are found in the various stages through which animals pass on their way from the egg to the fully formed animal. All animals above the protozoa begin their lives as single cells, and between that rudimentary condition and the adult stage every gradation of structure is exhibited. As animals de-

FIG. 107.—Reconstruction of the Ancestor of the Horse by Charles R. Knight, under the direction of Professor Osborn. Permission American Museum Natural History.

velop they become successively more and more complex, and in their shifting history many rudimentary organs arise

FIG. 108.—Fossil Remains of a Primitive Bird (Archeopteryx). From the specimen in the Berlin Museum. (After Kayser.)

and disappear. For illustration, in the young chick, developing within the hen's egg, there appear, after three or four days of incubation, gill-slits, or openings into the throat, like the gill-openings of lower fishes. These organs belong primarily to water life, and are not of direct use to the chick.

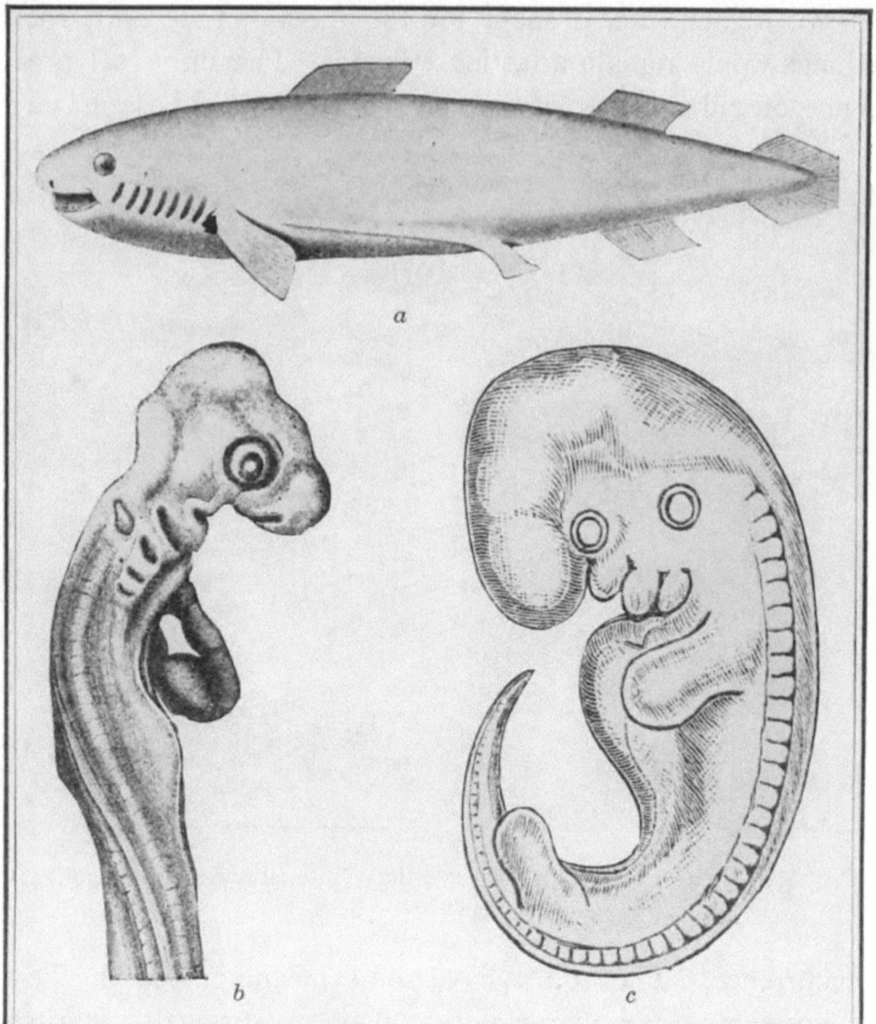

FIG. 109.—The Gill-clefts of a Shark (upper fig.) Compared with Those of the Embryonic Chick (to the left) and Rabbit.

The heart and the blood-vessels at this stage are also of the fish-like type, but this condition does not last long; the gill-slits, or gill-clefts, fade away within a few days, and the arteries of the head and the neck undergo great changes long before the chick is hatched. Similar gill-clefts and similar arrangements of blood-vessels appear also very early in the development of the young rabbit, and in the development of all higher life. Except for the theory of descent, such things would remain a lasting enigma. The universal presence of gill-clefts is not to be looked on as a haphazard

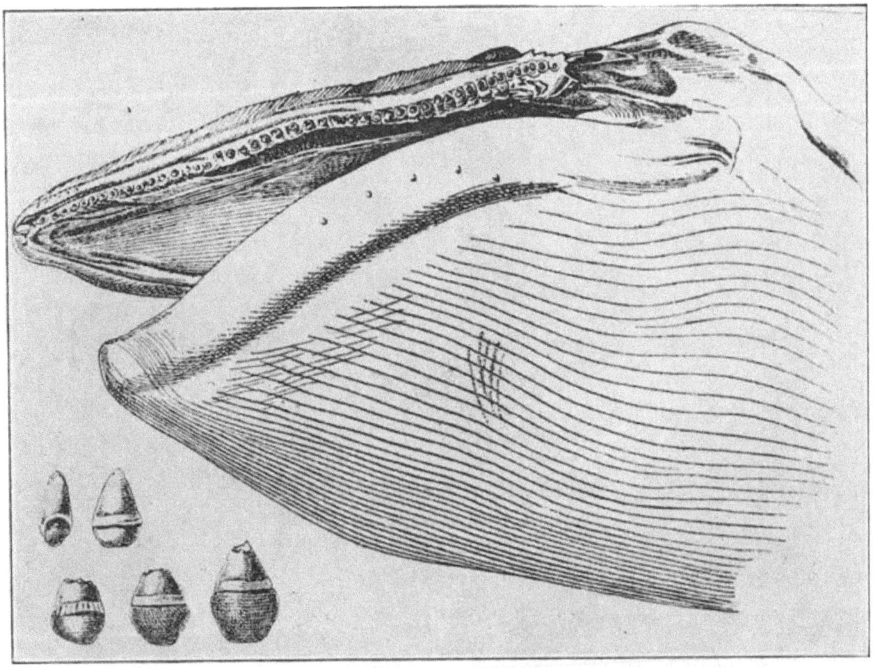

FIG. 110.—The Jaws of an Embryonic Whale, Showing Rudimentary Teeth.

occurrence. They must have some meaning, and the best suggestion so far offered is that they are survivals inherited from remote ancestors. The higher animals have sprung from simpler ones, and the gill-slits, along with other rudimentary organs, have been retained in their history. It is not necessary to assume that they are inherited from adult ancestors; they are, more likely, embryonic structures still retained in the developmental history of higher animals.

Such traces are like inscriptions on ancient columns—they are clues to former conditions, and, occurring in the animal series, they weigh heavily on the side of evolution.

An idea of the appearance of gill-clefts may be obtained from Fig. 109 showing the gill-clefts in a shark and those in the embryo of a chick and a rabbit.

Of a similar nature are the rudimentary teeth in the jaws of the embryo of the whalebone whale (Fig. 110). The adults have no teeth, these appearing only as transitory rudiments in the embryo. It is to be assumed that the teeth are inheritances, and that the toothless baleen whale is derived from toothed ancestors.

If we now turn to comparative anatomy, to classification, and to the geographical distribution of animals, we find that it is necessary to assume the doctrine of descent in order to explain the observed facts; the evidence for evolution, indeed, becomes cumulative. But it is not necessary, nor will space permit, to give extended illustrations from these various departments of biological researches.

The Human Body.—Although the broad doctrine of evolution rests largely upon the observation of animals and plants, there is naturally unusual interest as to its teaching in reference to the development of the human body. That the human body belongs to the animal series has long been admitted, and that it has arisen through a long series of changes is shown from a study of its structure and development. It retains marks of the scaffolding in its building. The human body has the same devious course of embryonic development as that of other mammals. In the course of its formation gill-clefts make their appearance; the circulation is successively that of a single-, a double-, and a four-chambered heart, with blood-vessels for the gill-clefts. Time and energy are consumed in building up rudimentary strue-tures which are evanescent and whose presence can be best explained on the assumption that they are, as in other animals, hereditary survivals.

Wiedersheim has pointed out more than one hundred and eighty rudimentary or vestigial structures belonging to the human body, which indicate an evolutionary relationship with lower vertebrates. It would require a considerable treatise to present the discoveries in reference to man's organization, as Wiedersheim has done in his *Structure of Man*. As passing illustrations of the nature of some of these suggestive things bearing on the question of man's origin may be mentioned: the strange grasping power of the newly born human infant, retained for a short time, and enabling the babe to sustain its weight; the presence of a tail and rudimentary tail muscles; of rudimentary ear muscles; of gill-clefts, etc.

Antiquity of Man.—The geological history of man is imperfectly known, although sporadic explorations have already accumulated an interesting series, especially as regards the shape and capacity of skulls. The remains of early quarternary man have been unearthed in various parts of Europe, and the probable existence of man in the tertiary period is generally admitted. As Osborn says, "Virtually three links have been found in the chain of human ancestry." The most primitive pre-human species is represented by portions of the skull and of the leg bones found in Java by the Dutch surgeon Dubois in the year 1890. These remains were found in tertiary deposits, and were baptized under the name of *Pithecanthropus erectus*. The structural position of this fossil is between the chimpanzee, the highest of anthropoid apes, and the "Neanderthal man." With characteristic scientific caution Osborn says that the *Pithecanthropus* "belongs in the line of none of the existing anthropoid apes, and falls very near, but not directly, in the line of human ancestry."

The second link is supplied by the famous Neanderthal skull found in the valley of the Neander, near Düsseldorf, in 1856. The discovery of this skull, with its receding forehead and prominent ridges above the orbits of the eyes, and its small cranial capacity, created a sensation, for it was soon seen that it was intermediate between the skulls of the lowest human races and those of the anthropoid apes. Virchow declared that if the skull was pre-human its structural characteristics were abnormal. This conclusion, however, was rendered untenable by the discovery in 1886 of similar skulls and the skeletons of two persons, in a cave near Spy in Bel-

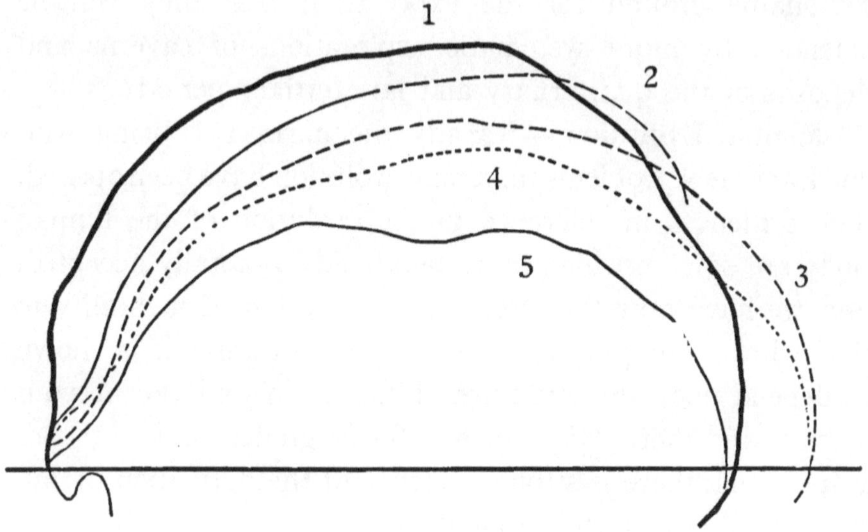

FIG. 111.—Profile Reconstructions of the Skulls of Living and Fossil Men: 1. Brachycephalic European; 2. The more ancient of the Nebraska skulls; 3. The Neanderthal man; 4. One of the Spy skulls; 5. Skull of the Java man. (Altered from Schwalbe and Osborn.)

gium. The "Spy man" and the "Neanderthal man" belong to the same type and are estimated to have been living in the middle of the palæolithic age.

The third link is in the early Neolithic man of Engis.

And now to this interesting series of gradations has been

added another by the discovery in 1906 of a supposed primitive race of men in Nebraska. The two skulls unearthed in Douglass County in that State indicate a cranial capacity falling below that of the "Australian negro, the lowest existing type of mankind known at present."

Fig. 111 shows in outline profile reconstructions of the skulls of some of the fossil types as compared with the short-headed type of Europe.

Palæontological discoveries are thus coming to support the evidences of man's evolution derived from embryology and archæology. While we must admit that the geological evidences are at present fragmentary, there is, nevertheless, reasonable ground for the expectation that they will be extended by more systematic explorations of caverns and deposits of the quarternary and late tertiary periods.

Mental Evolution.—Already the horizon is being widened, and new problems in human evolution have been opened. The evidences in reference to the evolution of the human body are so compelling as to be already generally accepted, and we have now the question of evolution of mentality to deal with. The progressive intelligence of animals is shown to depend upon the structure of the brain and the nervous system, and there exists such a finely graded series in this respect that there is strong evidence of the derivation of human faculties from brute faculties.

Sweep of the Doctrine of Evolution.—The great sweep of the doctrine of evolution makes it "one of the greatest acquisitions of human knowledge." There has been no point of intellectual vantage reached which is more inspiring. It is so comprehensive that it enters into all realms of thought. Weismann expresses the opinion that "the theory of descent is the most progressive step that has been taken in the development of human knowledge," and says that this position "is justified, it seems to me, even by this fact alone: that the evolution idea is not merely a new light on the special region of biological sciences, zoölogy and botany, but is of quite general importance. The conception of an evolution of life upon the earth reaches far beyond the bounds of any single science, and influences our whole realm of thought. It means nothing less than the elimination of the miraculous from our knowledge of nature, and the placing of the phenomena of life on the same plane as the other natural processes, that is, as having been brought about by the same forces and being subject to the same laws."

One feature of the doctrine is very interesting; it has enabled anatomists to predict that traces of certain structures not present in the adult will be found in the embryonic condition of higher animals, and by the verification of these predictions, it receives a high degree of plausibility. The presence of an *os centrale* in the human wrist was predicted, and afterward found, as also the presence of a rudimentary thirteenth rib in early stages of the human body. The predictions, of course, are chiefly technical, but they are based on the idea of common descent and adaptation.

It took a long time even for scientific men to arrive at a belief in the continuity of nature, and having arrived there, it is not easy to surrender it. There is no reason to think that the continuity is broken in the case of man's development. Naturalists have now come to accept as a mere statement of a fact of nature that the vast varietv of forms of life upon our globe has been produced bv a process of evolution. If this position be admitted, the next question would be, What are the factors which have been operative to bring this about? This brings us naturally to discuss the theories of evolution.

CHAPTER XVII

THEORIES OF EVOLUTION: LAMARCK, DARWIN

THE impression so generally entertained that the doctrine of organic evolution is a vague hypothesis, requiring for its support great stretches of the imagination, gives way to an examination of the facts, and we come to recognize that it is a well-founded theory, resting upon great accumulations of evidence. If the matter could rest here, it would be relatively simple; but it is necessary to examine into the causes of the evolutionary process. While scientific observation has shown that species are not fixed, but undergo transformations of considerable extent, there still remains to be accounted for the way in which these changes have been produced.

One may assume that the changes in animal life are the result of the interaction of protoplasm and certain natural agencies in its surroundings, but it is evidently a very difficult matter to designate the particular agencies or factors of evolution that have operated to bring about changes in species. The attempts to indicate these factors give rise to different theories of evolution, and it is just here that the controversies concerning the subject come in. We must remember, however, that to-day the controversies about evolution are not as to whether it was or was not the method of creation, but as to the factors by which the evolution of different forms was accomplished. Says Packard: "We are all evolutionists, though we may differ as to the nature of the efficient causes."

Of the various theories which had been advanced to

account for evolution, up to the announcement of the mutation-theory of De Vries in 1900, three in particular had commanded the greatest amount of attention and been the field for varied and extensive discussion. These are the theories of Lamarck, Darwin, and Weismann. They are comprehensive theories, dealing with the process as a whole. Most of the others are concerned with details, and emphasize certain phases of the process.

Doubtless the factors that have played a part in molding the forms that have appeared in the procession of life upon our globe have been numerous, and, in addition to those that have been indicated, Osborn very aptly suggests that there may be undiscovered factors of evolution. Within a few years De Vries has brought into prominence the idea of sudden transformations leading to new species, and has accounted for organic evolution on that basis. Further consideration of this theory, however, will be postponed, while in the present chapter we shall endeavor to bring out the salient features of the theories of Lamarck and Darwin, without going into much detail regarding them.

Lamarck

Lamarck was the first to give a theory of evolution that has retained a place in the intellectual world up to the present time, and he may justly be regarded as the founder of that doctrine in the modern sense. The earlier theories were more restricted in their reach than that of Lamarck. Erasmus Darwin, his greatest predecessor in this field of thought, announced a comprehensive theory, which, while suggestive and forceful in originality, was diffuse, and is now only of historical importance. The more prominent writers on evolution in the period prior to Lamarck will be dealt with in the chapter on the Rise of Evolutionary Thought.

Lamarck was born in 1744, and led a quiet, monotonous life, almost pathetic on account of his struggles with poverty, and the lack of encouragement and proper recognition by his contemporaries. His life was rendered more bearable, however, even after he was overtaken by complete blindness, by the intellectual atmosphere that he created for himself, and by the superb confidence and affection of his devoted daughter Cornélie, who sustained him and made the truthful prediction that he would be recognized by posterity (" *La postérité vous honorera*").

His Family.—He came of a military family possessing some claims to distinction. The older name of the family had been de Monet, but in the branch to which Lamarck belonged the name had been changed to de Lamarque, and in the days of the first Republic was signed plain Lamarck by the subject of this sketch. Jean Baptiste Lamarck was the eleventh and last child of his parents. The other male members of the family having been provided with military occupations, Jean was selected by his father, although against the lad's own wish, for the clerical profession, and accordingly was placed in the college of the Jesuits at Amiens. He did not, however, develop a taste for theological studies, and after the death of his father in 1760 "nothing could induce the incipient abbé, then seventeen years of age, longer to wear his bands."

His ancestry asserted itself, and he forsook the college to follow the French army that was then campaigning in Germany. Mounted on a broken-down horse which he had succeeded in buying with his scanty means, he arrived on the scene of action, a veritable raw recruit, appearing before Colonel Lastic, to whom he had brought a letter of recommendation.

Military Experience.—The Colonel would have liked to be rid of him, but owing to Lamarck's persistence, assigned him to a company; and, being mounted, Lamarck took rank as a sergeant. During his first engagement his company was exposed to the direct fire of the enemy, and the officers one after another were shot until Lamarck by order of succession was in command of the fourteen remaining grenadiers. Although the French army retreated, Lamarck refused to move with his squad until he received directions from headquarters to retire. In this his first battle he showed the courage and the independence that characterized him in later years.

Adopts Natural Science.—An injury to the glands of the neck, resulting from being lifted by the head in sport by one of his comrades, unfitted him for military life, and he went to Paris and began the study of medicine, supporting himself in the mean time by working as a bank clerk. It was in his medical course of four years' severe study that Lamarck received the exact training that was needed to convert his enthusiastic love for science into the working powers of an investigator. He became especially interested in botany, and, after a chance interview with Rousseau, he determined to follow the ruling passion of his nature and devote himself to natural science. After about nine years' work he published, in 1778, his *Flora of France*, and in due course was appointed to a post in botany in the Academy of Sciences. He did not hold this position long, but left it to travel with the sons of Buffon as their instructor. This agreeable occupation extended over two years, and he then returned to Paris, and soon after was made keeper of the herbarium in the Royal Garden, a subordinate position entirely beneath his merits. Lamarck held this poorly paid position for several years, and was finally relieved by being appointed a professor in the newly established *Jardin des Plantes*.

He took an active part in the reorganization of the Royal Garden (*Jardin du Roi*) into the *Jardin des Plantes*. When, during the French Revolution, everything that was suggestive of royalty became obnoxious to the people, it was Lamarck who suggested in 1790 that the name of the King's Garden be changed to that of the Botanical Garden (*Jardin des Plantes*). The Royal Garden and the Cabinet of Natural History were combined, and in 1793 the name Jardin des Plantes proposed by Lamarck was adopted for the institution.

It was through the endorsements of Lamarck and Geoffroy Saint-Hilaire that Cuvier was brought into this great scientific institution; Cuvier, who was later to be advanced above him in the Jardin and in public favor, and who was to break friendship with Lamarck and become the opponent of his views, and who also was to engage in a memorable debate with his other supporter, Saint-Hilaire.

The portrait of Lamarck shown in Fig. 112 is one not generally known. Its date is undetermined, but since it was published in Thornton's *British Plants* in 1805, we know that it was painted before the publication of Lamarck's *Philosophie Zoologique*, and before the full force of the coldness and heartless neglect of the world had been experienced. In his features we read supremacy of the intellect, and the unflinching moral courage for which he was notable. Lamarck has a more hopeful expression in this portrait than in those of his later years.

Lamarck Changes from Botany to Zoölogy.—Until 1794, when he was fifty years of age, Lamarck was devoted to botany, but on being urged, after the reorganization of the *Jardin du Roi*, to take charge of the department of invertebrates, he finally consented and changed from the study of plants to that of animals. This change had profound influence in shaping his ideas. He found the invertebrates in great confusion, and set about to bring order out of chaos, an undertaking in which, to his credit be it acknowledged, he succeeded. The fruit of his labors, the Natural History of Invertebrated Animals (*Historie naturelle des Ani-*

FIG. 112.—LAMARCK, 1774–1829.
From Thornton's *British Plants*, 1805.

maux sans Vertèbres, 1815–1822), became a work of great importance. He took hold of this work, it should be remembered, as an expert observer, trained to rigid analysis by his previous critical studies in botany. In the progress of the work he was impressed with the differences in animals and the difficulty of separating one species from another. He had occasion to observe the variations produced in animals through the influence of climate, temperature, moisture, elevation above the sea-level, etc.

He observed also the effects of use and disuse upon the development of organs: the exercise of an organ leading to its greater development, and the disuse to its degeneration. Numerous illustrations are cited by Lamarck which serve to make his meaning clear. The long legs of wading birds are produced and extended by stretching to keep above the water; the long neck and bill of storks are produced by their habit of life; the long neck of the giraffe is due to reaching for foliage on trees; the web-footed birds, by spreading the toes when they strike the water, have stimulated the development of a membrane between the toes, etc. In the reverse direction, the loss of the power of flight in the "wingless" bird of New Zealand is due to disuse of the wings; while the loss of sight in the mole and in blind cave animals has arisen from lack of use of eves.

The changes produced in animal organization in this way were believed to be continued bv direct inheritance and improved in succeeding generations.

He believed also in a perfecting principle, tending to improve animals—a sort of conscious endeavor on the part of the animal playing a part in its better development. Finally, he came to believe that the agencies indicated above were the factors of the evolution of life.

His Theory of Evolution.—All that Lamarck had written before he changed from botany to zoölogy (1794) indicates his belief in the fixity of species, which was the prevailing notion among naturalists of the period. Then, in 1800, we find him apparently all at once expressing a contrary opinion, and an opinion to which he held unwaveringly to the close of his life. It would be of great interest to determine when Lamarck changed his views, and upon what this radical reversal of opinion was based; but we have no sure record to depend upon. Since his theory is developed chiefly upon considerations of animal life, it is reasonable to assume that his evolutionary ideas took form in his mind after he began the serious study of animals. Doubtless, his mind having been prepared and his insight sharpened by his earlier studies, his observations in a new field supplied the data which led him directly to the conviction that species are unstable. As Packard, one of his recent biographers, points out, the first expression of his new views of which we have any record occurred in the spring of 1800, on the occasion of his opening lecture to his course on the invertebrates. This avowal of belief in the extensive alteration of species was published in 1801 as the preface to his *Système des Animaux sans Vertèbres*. Here also he foreshadowed his theory of evolution, saying that nature, having formed the simplest organisms, "then with the aid of much time and favorable circumstances . . . formed all the others." It has been generally believed that Lamarck's first public expression of his views on evolution was published in 1802 in his *Recherches sur l'Organisation des Corps Vivans*, but the researches of Packard and others have established the earlier date.

Lamarck continued for several years to modify and amplify the expression of his views. It is not necessary, however, to follow the molding of his ideas on evolution as expressed in the opening lectures to his course in the years 1800, 1802, 1803, and 1806, since we find them fully elaborated in his *Philosophie Zoologique*, published in 1809, and this may be accepted as the standard source for the study of his theory. In this work he states two propositions under the name of laws, which have been translated by Packard as follows:

"*First Law:* In every animal which has not exceeded the term of its development, the more frequent and sustained use of any organ graduallv strengthens this organ, develops and enlarges it, and gives it a strength proportioned to the length of time of such use; while the constant lack of use of such an organ imperceptibly weakens it, causing it to become reduced, progressively diminishes its faculties, and ends in its disappearance.

"*Sccond Law:* Everything which nature has caused individuals to acquire or lose by the influence of the circumstances to which their race may be for a long time exposed, and consequently by the influence of the predominant use of such an organ, or by that of the constant lack of use of such part, it preserves by hereditv and passes on to the new individuals which descend from it, provided that the changes thus acquired are common to both sexes, or to those which have given origin to these new individuals.

"These are the two fundamental truths which can be misunderstood only by those who have never observed or followed nature in its operations," etc. The first law embodies the principle of use and disuse, the second law that of heredity.

In 1815 his theory received some extensions of minor importance. The only points to which attention need be called are that he gives four laws instead of two, and that a new feature occurs in the second law in the statement that the production of a new organ is the result of a new need (*besoin*) which continues to make itself felt.

Simplified Statement of Lamarck's Views.—For practical exposition the theory may be simplified into two sets of facts: First, those to be classed under variation; and, second, those under heredity. Variations of organs, according to Lamarck, arise in animals mainly through use and disuse, and new organs have their origin in a physiological need. A new need felt by the animal expresses itself on the organism, stimulating growth and adaptations in a particular direction. This part of Lamarck's theory has been subjected to much ridicule. The sense in which he employs the word *besoin* has been much misunderstood; when, however, we take into account that he uses it, not merely as expressing a wish or desire on the part of the animal, but as the reflex action arising from new conditions, his statement loses its alleged grotesqueness and seems to be founded on sound physiology.

Inheritance.—Lamarck's view of heredity was uncritical; according to his conception, inheritance was a simple, direct transmission of those superficial changes that arise in organs within the lifetime of an individual owing to use and disuse. It is on this question of the direct inheritance of variations acquired in the lifetime of an individual that his theory has been the most assailed. The belief in the inheritance of acquired characteristics has been so undermined by experimental evidence that at the present time we can not point to a single unchallenged instance of such inheritance. But, while Lamarck's theory has shown weakness on that side, his ideas regarding the production of variations have been revived and extended.

Variation.—The more commendable part of his theory is the attempt to account for variation. Darwin assumed variation, but Lamarck attempted to account for it, and in this feature many discerning students maintain that the theory of Lamarck is more philosophical in its foundation than that of Darwin.

In any theory of evolution we must deal with the variation of organisms and heredity, and thus we observe that the two factors discussed by Lamarck are basal. Although it must be admitted that even to-day we know little about either variation or heredity, they remain basal factors in any theory of evolution.

Time and Favorable Conditions.—Lamarck supposed a very long time was necessary to bring about the changes which have taken place in animals. The central thought of time and favorable conditions occurs again and again in his writings. The following quotation is interesting as coming from the first announcement of his views in 1800:

"It appears, as I have already said, that *time* and *favorable conditions* are the two principal means which nature has employed in giving existence to all her productions. We know that for her time has no limit, and that consequently she has it always at her disposal.

"As to the circumstances of which she has had need and of which she makes use every day in order to cause her productions to vary, we can say that in a manner they are inexhaustible.

"The essential ones arising from the influence and from all the environing media, from the diversity of local causes, of habits, of movements, of action, finally of means of living, of preserving their lives, of defending themselves, of multiplying themselves, etc. Moreover, as the result of these different influences, the faculties, developed and strengthened by use, become diversified by the new habits maintained for long ages, and by slow degrees the structure, the consistence—in a word, the nature, the condition of the parts and of the organs consequently participating in all these influences, became preserved and were propagated by heredity (génération)." (Packard's translation.)

Salient Points.—The salient points in Lamarck's theory may be compacted into a single sentence: It is a theory of the evolution of animal life, depending upon variations brought about mainly through use and disuse of parts, and also by responses to external stimuli, and the direct inheritance of the same. His theory is comprehensive, so much so that he includes mankind in his general conclusions.

Lamarck supposed that an animal having become adapted to its surroundings would remain relatively stable as to its structure. To the objection raised by Cuvier that animals from Egypt had not changed since the days when they were preserved as mummies, he replied that the climate of Egypt had remained constant for centuries, and therefore no change in its fauna was to be expected.

Species.—Since the question of the fixity of species is the central one in theories of evolution, it will be worth while to quote Lamarck's definition of species: "All those who have had much to do with the study of natural history know that naturalists at the present day are extremely embarrassed in defining what they mean by the word species. . . . We call *species* every collection of individuals which are alike or almost so, and we remark that the regeneration of these individuals conserves the species and propagates it in continuing successively to reproduce similar individuals." He then goes on with a long discussion to show that large collections of animals exhibit a great variation in species, and that they have no absolute stability, but "enjoy only a relative stability."

Herbert Spencer adopted and elaborated the theory of Lamarck. He freed it from some of its chief crudities, such as the idea of an innate tendency toward perfection. In many controversies Mr. Spencer defended the idea of the transmission of acquired characters. The ideas of Lamarck have, therefore, been transmitted to us largely in the Spencerian mold and in the characteristic language of that great philosopher. There has been but little tendency to go to Lamarck's original writings. Packard, whose biography of Lamarck appeared in 1901, has made a thorough analysis of his, writings and had incidentally corrected several erroneous conception.

Neo-Lamarckism.—The ideas of Lamarck regarding the beginning of variations have been revived and accorded much respect under the designation of Neo-Lamarckism. The revival of Lamarckism is especially owing to the palæontological investigations of Cope and Hyatt. The work of E. D. Cope in particular led him to attach importance to the effect of mechanical and other external causes in producing variation, and he points out many instances of use-inheritance. Neo-Lamarckism has a considerable following; it is a revival of the fundamental ideas of Lamarck.

Darwin's Theory

While Lamarck's theory rests upon two sets of facts, Darwin's is founded on three: *viz.*, the facts of variation, of inheritance, and of natural selection. The central feature of his theory is the idea of natural selection. No one else save Wallace had seized upon this feature when Darwin made it the center of his system. On account of the part taken by Wallace simultaneously with Darwin in announcing natural selection as the chief factor of evolution, it is appropriate to designate this contribution as the Darwin-Wallace principle of natural selection. The interesting connection between the original conclusions of Darwin and Wallace is set forth in Chapter XIX.

Variation.—It will be noticed that two of the causes assigned by Darwin are the same as those designated by Lamarck, but their treatment is quite different. Darwin (Fig. 113) assumed variation among animals and plants without attempting to account for it, while Lamarck undertook to state the particular influences which produce variation, and although we must admit that Lamarck was not entirely suc-

Fig. 113.—Charles Darwin, 1809–1882.

cessful in this attempt, the fact that he undertook the task places his contribution at the outset on a very high plane.

The existence of variation as established by observation is unquestioned. No two living organisms are exactly alike at the time of their birth, and even if they are brought up together under identical surroundings they vary. The variation of plants and animals under domestication is so conspicuous and well known that this kind of variation was the first to attract attention. It was asserted that these variations were perpetuated because the forms had been protected by man, and it was doubted that animals varied to any considerable extent in a state of nature. Extended collections and observations in field and forest have, however, set this question at rest.

If crows or robins or other birds are collected on an extensive scale, the variability of the same species will be evident Many examples show that the so-called species differ greatly in widely separated geographical areas, but collections from the intermediate territory demonstrate that the variations are connected by a series of fine gradations. If, for illustration, one should pass across the United States from the Atlantic to the Pacific coast, collecting one species of bird, the entire collection would exhibit wide variations, but the extremes would be connected by intermediate forms.

The amount of variation in a state of nature is much greater than was at first supposed, because extensive colleetions were lacking, but the existence of wide variation is now established on the basis of observation. This fact of variation among animals and plants in the state of nature is unchallenged, and affords a good point to start from in con sidering Darwinism.

Inheritance.—The idea that these variations are inherited is the second point. But what particular variations will be preserved and fostered by inheritance, and on what principle they will be selected, is another question—and a notable one. Darwin's reply was that those variations which are of advantage to the individual will be the particular ones selected by nature for inheritance. While Darwin implies the inheritance of acquired characteristics, his theory of heredity was widely different from that of Lamarck. Darwin's theory of heredity, designated the provisional theory of pangenesis, has been already considered (see Chapter XIV).

Natural Selection.—Since natural selection is the main feature of Darwin's doctrine, we must devote more time to it. Darwin frequently complained that very few of his critics took the trouble to find out what he meant by the term natural selection. A few illustrations will make his meaning clear. Let us first think of artificial selection as it is applied by breeders of cattle, fanciers of pigeons and of other fowls, etc. It is well known that by selecting particular variations in animals and plants, even when the variations are slight, the breeder or the horticulturalist will be able in a short time to produce new races of organic forms. This artificial selection on the part of man has given rise to the various breeds of dogs, the 150 different kinds of pigeons, etc., all of which breed true. The critical question is, Have these all an individual ancestral form in nature? Observation shows that many different kinds—as pigeons—may be traced back to a single ancestral form, and thus the doctrine of the fixity of species is overthrown.

Now, since it is demonstrated by observation that variations occur, if there be a selective principle at work in nature, effects similar to those caused by artificial selection will be produced. The selection by nature of the forms fittest to survive is what Darwin meant by natural selection. We can never understand the application, however, unless we take into account the fact that while animals tend to multiply in geometrical progression, as a matter of fact the number of any one kind remains practically constant. Although the face of nature seems undisturbed, there is nevertheless a struggle for existence among all animals.

This is easily illustrated when we take into account the breeding of fishes. The trout, for illustration, lays from 60,000 to 100,000 eggs. If the majority of these arrived at maturity and gave rise to progeny, the next generation would represent a prodigious number, and the numbers in the succeeding generations would increase so rapidly that soon there would not be room in the fresh waters of the earth to contain their descendants. What becomes of the immense number of fishes that die? They fall a prey to others, or they are not able to get food in competition with other more hardy relatives, so that it is not a matter of chance that determines which ones shall survive; those which are the strongest, the better fitted to their surroundings, are the ones which will be perpetuated.

The recognition of this struggle for existence in nature, and the consequent survival of the fittest, shows us more clearly what is meant by natural selection. Instead of man making the selection of those particular forms that are to survive, it is accomplished in the course of nature. This is natural selection.

Various Aspects of Natural Selection.—Further illustrations are needed to give some idea of the various phases of natural selection. Speed in such animals as antelopes may be the particular thing which leads to their protection. It stands to reason that those with the greatest speed would escape more readily from their enemies, and would be the particular ones to survive, while the weaker and slower ones would fall victims to their prey. In all kinds of strain due to scarcity of food, inclemency of weather, and other untoward circumstances, the forms which are the strongest, physiologically speaking, will have the best chance to weather the strain and to survive. As another illustration, Darwin pointed out that natural selection had produced a long-legged race of prairie wolves, while the timber wolves, which have less occasion for running, are short-legged.

We can also see the operation of natural selection in the production of the sharp eyes of birds of prey. Let us consider the way in which the eyes of the hawk have been perfected by evolution. Natural selection compels the eye to come up to a certain standard. Those hawks that are born with weak or defective vision cannot cope with the conditions under which they get their food. The sharp-eyed forms would be the first to discern their prey, and the most sure in seizing upon it. Therefore, those with defective vision or with vision that falls below the standard will be at a very great disadvantage. The sharp-eyed forms will be preserved by a selective process. Nature selects, we may say, the keener-eyed birds of prey for survival, and it is easy to see that this process of natural selection would establish and maintain a standard of vision.

But natural selection tends merelv to adapt animals to their surroundings, and does not always operate in the direc tion of increasing the efficiency of the organ. We take another illustration to show how Darwin explains the origin of races of short-winged beetles on certain oceanic islands. Madeira and other islands, as Kerguelen island of the Indian Ocean, are among the most windv places in the world. The strong-winged beetles, being accustomed to disport themselves in the air, would be carried out to sea by the sudden and violent gales which sweep over those islands, while the weaker-winged forms would be left to perpetuate their kind. Thus, generation after generation, the strong-winged beetles would be eliminated by a process of natural selection, and there would be left a race of short-winged beetles derived from long-winged ancestors. In this case the organs are reduced in their development, rather than increased; but manifestly the short-winged race of beetles is better adapted to live under the particular conditions that surround their life in these islands.

While this is not a case of increase in the particular organ, it illustrates a progressive series of steps whereby the organism becomes better adapted to its surroundings. A similar instance is found in the suppression of certain sets of organs in internal parasites. For illustration, the tapeworm loses particular organs of digestion for which it does not have continued use; but the reproductive organs, upon which the continuance of its life depends, are greatly increased. Such cases as the formation of short-winged beetles show us that the action of natural selection is not always to preserve what we should call the best, but simply to preserve the fittest. Development, therefore, under the guidance of natural selection is not always progressive. Selection by nature does not mean the formation and preservation of the ideally perfect, but merely the survival of those best fitted to their environment.

Color.—The various ways in which natural selection acts are exceedingly diversified. The colors of animals may be a factor in their preservation, as the stripes on the zebra tending to make it inconspicuous in its surroundings. The stripes upon the sides of tigers simulate the shadows cast by the jungle grass in which the animals live, and serves to conceal them from their prey as well as from enemies. Those animals that assume a white color in winter become thereby less conspicuous, and they are protected by their coloration.

As further illustrating color as a factor in the preservation of animals, we may cite a story originally told by Professor E. S. Morse. When he was collecting shells on the white sand of the Japanese coast, he noticed numerous white tiger-beetles, which could scarcely be seen against the white background. They could be detected chiefly by their shadows when the sun was shining. As he walked along the coast he came to a wide band of lava which had flowed from a crater across the intervening countrv and plunged into the sea, leaving a broad dark band some miles in breadth across the white sandy beach. As he passed from the white sand to the dark lava, his attention was attracted to a tiger-beetle almost identical with the white one except as to color. Instead of being white, it was black. He found this broad, black band of lava inhabited by the black tiger beetle, and found very few, if any, of the white kind. This is a striking illustration of what has occurred in nature. These two beetles are of the same species, and in examining the conditions under which they grow, it is discovered that out of the eggs laid bv the original white forms, there now and then appears one of a dusky or black color. Consider how conspicuous this dark object would be against the white background of sand. It would be an easv mark for the birds of prey that fly about, and therefore on the white surface the black beetles would be destroyed, while the white ones would be left. But on the black background of lava the conditions are reversed. There the white forms would be the conspicuous ones; as they wandered upon the black surface, thev would be picked up by birds of prey and the black ones would be left. Thus we see another instance of the operation of natural selection.

Mimicry.—We have, likewise, in nature a great number of cases that are designated mimicry. For illustration, certain caterpillars assume a stiff position, resembling a twig from a branch. We have also leaf-like butterflies. The Kallima of India is a conspicuous illustration of a butterfly having the upper surface of its wings bright-colored, and the lower surface dull. When it settles upon a twig the wings are closed and the under-sides have a mark across them resembling the mid-rib of a leaf, so that the whole butterfly in the resting position becomes inconspicuous, being protected by mimicry.

One can readily see how natural selection would be evoked in order to explain this condition of affairs. Those forms that varied in the direction of looking like a leaf would be the most perfectly protected, and this feature being fostered by natural selection, would, in the course of time, produce a race of butterflies the resemblance of whose folded wings to a leaf would serve as a protection from enemies.

It may not be out of place to remind the reader that the illustrations cited are introduced merely to elucidate Darwin's theory and the writer is not committed to accepting them as explanations of the phenomena involved. He is not unmindful of the force of the criticisms against the adequacy of natural selection to explain the evolution of all kinds of organic structures.

Many other instances of the action of color might be added, such as the wearing of warning colors, those colors which belong to butterflies, grubs, and other animals that have a noxious taste. These warning colors have taught birds to leave alone the forms possessing those colors. Sometimes forms which do not possess a disagreeable taste secure protection by mimicking the colors of the noxious varieties.

Sexual Selection.—There is an entirely different set of cases which at first sight would seem difficult to explain on the principle of selection. How, for instance, could we explain the feathers in the tails of the birds of paradise, or that peculiar arrangement of feathers in the tail of the lyre-bird, or the gorgeous display of tail-feathers of the male peacock? Here Mr. Darwin seized upon a selective principle arising from the influence of mating. The male birds in becoming suitors for a particular female have been accustomed to display their tail-feathers; the one with the most attractive display excites the pairing instinct in the highest degree, and becomes the selected suitor. In this way, through the operation of a form of selection which Darwin designates sexual selection, possibly such curious adaptations as the peacock's tail may be accounted for.

It should be pointed out that this part of the theory is almost wholly discredited by biologists. Experimental evidence is against it. Nevertheless in a descriptive account of Darwin's theory it may be allowed to stand without critical comment.

Inadequacy of Natural Selection.—In nature, under the struggle for existence, the fittest will be preserved; and natural selection will operate toward the elaboration or the suppression of certain organs or certain characteristics when the elaboration or the suppression is of advantage to the animal form. Much has been said of late as to the inadequacy of natural selection. Herbert Spencer and Huxley, both accepting natural selection as one of the factors, doubted its complete adequacy.

One point is often overlooked, and should be brought out with clearness; *viz.*, that Darwin himself was the first to point out clearly the inadequacy of natural selection as a universal law for the production of the great variety of animals and plants. In the second edition of the *Origin of Species* he says: "But, as my conclusions have lately been much misrepresented, and it has been stated that I attribute the modification of species exclusively to natural selection, I may be permitted to remark that in the first edition of this work and subsequently I placed in a most conspicuous position,—namely, at the close of the introduction—the following words: 'I am convinced that natural selection has been the main, but not the exclusive means of modification.' This has been of no avail. Great is the power of steady misrepresentation. But the history of science shows that fortunately this power does not long endure."

The reaction against the all-sufficiency of natural selection, therefore, is something which was anticipated by Darwin, and the quotation made above will be a novelty to many of our readers who supposed that they understood Darwin's position.

Confusion between Lamarck's and Darwin's Theories.—Besides the failure to understand what Darwin has written, there is great confusion, both in pictures and in writings, in reference to the theories of Darwin and Lamarck. Poulton illustrated a state of confusion in one of his lectures on the theory of organic evolution, and the following instances are quoted from memory.

We are most of us familiar with such pictures as the following: A man standing and waving his arms; in the next picture these arms and hands become enlarged, and in the successive pictures they undergo transformations into wings, and the transference is made into a flying animal.

Such pictures are designated "The origin of flight after Darwin." The interesting circumstance is this, that the illustration does not apply to Darwin's idea of natural selection at all, but is pure Lamarckism. Lamarck contended for the production of new organs through the influence of use and disuse, and this particular illustration refers to that, and not to natural selection at all.

Among the examples of ridicule to which Darwin's ideas have been exposed, we cite one verse from the song of Lord Neaves. His lordship wrote a song with a large number of verses hitting off in jocular vein many of the claims and foibles of his time. In attempting to make fun of Darwin's idea he misses completely the idea of natural selection, but hits upon the principle enunciated by Lamarck, instead. He says:

"A deer with a neck which was longer by half
Than the rest of his family's—try not to laugh—
By stretching and stretching became a giraffe,
Which nobody can deny."

The clever young woman, Miss Kendall, however in her *Song of the Ichthyosaurus*, showed clearness in grasping Darwin's idea when she wrote:

"Ere man was developed, our brother
We swam, we ducked, and we dived,
And we dined, as a rule, on each other.
What matter? The toughest survived."

This hits the idea of natural selection. The other two illustrations miss it, but strike the principle which was enunciated by Lamarck. This confusion between Lamarckism and Darwinism is very wide-spread.

Darwin's book on the *Origin of Species*, published in 1859, was epoch-making. If a group of scholars were asked to designate the greatest book of the nineteenth century—that is, the book which created the greatest intellectual stir—it is likely that a large proportion of them would reply that it is Darwin's *Origin of Species.* Its influence was so great in the different domains of thought that we may observe a natural cleavage between the thought in reference to nature between 1859 and all preceding time. His other less widely known books on *Animals and Plants Under Domestication*, the *Descent of Man*, etc., etc., are also important contributions to the discussion of his theory. A brief account of Darwin, the man, will be found in Chapter XIX.

CHAPTER XVIII

THEORIES OF EVOLUTION CONTINUED: WEISMANN, DE VRIES

WEISMANN'S views have passed through various stages of remodeling since his first public championship of the Theory of Descent on assuming, in 1867, the position of professor of zoölogy in the University of Freiburg. Some time after that date he originated his now famous theory of heredity, which has been retouched, from time to time, as the result of aggressive criticism from others, and the expansion of his own mental horizon. As he said in 1904, regarding his lectures on evolution which have been delivered almost regularly every year since 1880, they "were gradually modified in accordance with the state of my knowledge at the time, so that they have been, I may say, a mirror of my own intellectual evolution."

Passing over his book, *The Germ Plasm*, published in English in 1893, we may fairly take his last book, *The Evolution Theory*, 1904, as the best exposition of his conelusions. The theoretical views of Weismann have been the field of so much strenuous controversy that it will be well perhaps to take note of the spirit in which they have been presented. In the preface of his book just mentioned, he says: "I make this attempt to sum up and present as a harmonious whole the theories which for forty years I have been gradually building up on the basis of the legacy of the great workers of the past, and on the results of my own investigations and those of my fellow-workers, not because I regard the picture as incomplete or incapable of improvement, but because I believe its essential features to be correct, and because an eye-trouble which has hindered my work for many years makes it uncertain whether I shall have much more time and strength granted to me for its further elaboration."

The germ-plasm theory is primarily a theory of heredity, and only when connected with other considerations does it become the full-fledged theory of evolution known as Weismannism. The theory as a whole involves so many intricate details that it is difficult to make a clear statement of it for general readers. If in considering the theories of Lamarck and Darwin it was found advantageous to confine attention to salient points and to omit details, it is all the more essential to do so in the discussion of Weismann's theory.

In his prefatory note to the English edition of *The Evolution Theory* Thomson, the translator, summarizes Weismann's especial contributions as: "(1) the illumination of the evolution process with a wealth of fresh illustrations; (2) the vindication of the 'germ-plasm' concept as a valuable working hypothesis; (3) the final abandonment of any assumption of transmissible acquired characters; (4) a further analysis of the nature and origin of variations; and (5), above all, an extension of the selection principle of Darwin and Wallace, which finds its logical outcome in the suggestive theory of germinal selection."

Continuity of the Germ-Plasm.—Weismann's theory is designated that of continuity of the germ-plasm, and in considering it we must first give attention to his conception of the germ-plasm. As is well known, animals and plants spring from germinal elements of microscopic size; these are, in plants, the spores, the seeds, and their fertilizing agents; and, in animals, the eggs and the sperms. Now, since all animals, even the highest developed, begin in a fertilized egg, that structure, minute as it is, must contain all hereditary qualities, since it is the only material substance that passes from one generation to another. This hereditary substance is the germ-plasm. It is the living, vital substance of organisms that takes part in the development of new generations.

Naturalists are agreed on this point, that the more complex animals and plants have been derived from the simpler ones; and, this being accepted, the attention should be fixed on the nature of the connection between generations during their long line of descent. In the reproduction of single-celled organisms, the substance of the entire body is divided during the transmission of life, and the problem both of heredity and origin is relatively simple. It is clear that in these single-celled creatures there is unbroken continuity of body-substance from generation to generation. But in the higher animals only a minute portion of the organism is passed along.

Weismann points out that the many-celled body was gradually produced by evolution; and that in the transmission of life by the higher animals the continuitv is not between body-cells and their like, but only between ger minal elements around which in due course new body-cells are developed. Thus he regards the body-cells as constituting a sort of vehicle within which the germ-cells are carried. The germinal elements represent the primordial substance around which the body has been developed, and since in all the long process of evolution the germinal elements have been the only form of connection between different generations, thev have an unbroken continuity.

This conception of the continuity of the germ-plasm is the foundation of Weismann's doctrine. As indicated before, the general way in which he accounts for heredity is that the offspring is like the parent because it is composed of some of the same stuff. The rise of the idea of germinal continuity has been indicated in Chapter XIV, where it was pointed out that Weismann was not the originator of the idea, but he is nevertheless the one who has developed it the most extensively.

Complexity of the Germ-Plasm.—The germ-plasm has been molded for so many centuries by external circumstances that it has acquired an organization of great complexity. This appears from the following considerations: Protoplasm is impressionable; in fact, its most characteristic feature is that it responds to stimulation and modifies itself accordingly. These subtle changes occurring within the protoplasm affect its organization, and in the long run it is the summation of experiences that determines what the protoplasm shall be and how it will behave in development. Two masses of protoplasm differ in capabilities and potentialities according to the experiences through which they have passed, and no two will be absolutely identical. All the time the body was being evolved the protoplasm of the germinal elements was being molded and changed, and these ele ments therefore possess an inherited orgnization of great complexity.

When the body is built anew from the germinal elements, the derived qualities come into play, and the whole process is a succession of responses to stimulation. This is in a sense, on the part of the protoplasm, a repeating of its historical experience. In building the organism it does not go over the ground for the first time, but repeats the activities which it took centuries to acquire.

The evident complexity of the germ-plasm made it necessary for Weismann, in attempting to explain inheritance in detail, to assume the existence of distinct vital units within the protoplasm of the germinal elements. He has invented names for these particular units as biophors, the elementary vital units, and their combination into determinants, the latter being united into ids, idants, etc. The way in which he assumes the interactions of these units gives to his theory a highly speculative character. The conception of the complex organization of the germ-plasm which Weismann reached on theoretical grounds is now being established on the basis of observation (see Chapter XIV, p. 313).

The Origin of Variations.—The way in which Weismann accounts for the origin of variation among higher animals is both ingenious and interesting. In all higher organisms the sexes are separate, and the reproduction of their kind is a sexual process. The germinal elements involved are seeds and pollen, eggs and sperms. In animals the egg bears all the hereditary qualities from the maternal side, and the sperm those from the paternal side. The intimate mixture of these in fertilization gives great possibilities of variations arising from the different combinations and permutations of the vital units within the germ-plasm.

This union of two germ-plasms Weismann calls amphimixis, and for a long time he maintained that the purpose of sexual reproduction in nature is to give origin to variations. Later he extended his idea to include a selection, mainly on the basis of nutrition, among the vital elements composing the germ-plasm. This is germinal selection, which aids in the production of variations.

In *The Evolution Theory*, volume II, page 196, he says: "Now that I understand these processes more clearly, my opinion is that the roots of all heritable variation lie in the germ-plasm; and, furthermore, that the determinants are continually oscillating hither and thither in response to very minute nutritive changes and are readily compelled to *variation in a definite direction*, which may ultimately lead to considerable variations in the structure of the species, if they are favored by personal selection, or at least if they are not suppressed by it as prejudicial."

But while sexual reproduction may be evoked to explain the origin of variation in higher animals, Weismann thought it was not applicable to the lower ones, and he found himself driven to assume that variation in single-celled organisms is owing to the direct influence of environment upon them, and thus he had an awkward assumption of variations arising in a different manner in the higher and in the simplest organisms. If I correctly understand his present position, the conception of variation as due to the direct influence of environment is being surrendered in favor of the action of germinal selection among the simplest organisms.

Extension of the Principle of Natural Selection.—These variations, once started, will be fostered by natural selection provided they are of advantage to the organism in its struggle for existence. It should be pointed out that Weismann is a consistent Darwinian; he not only adopts the principle of natural selection, but he extends the field of its operation from externals to the internal parts of the germinal elements.

"Roux and others have elaborated the idea of a struggle of the parts within the organism, and of a corresponding intra-selection; . . . but Weismann, after his manner, has carried the selection-idea a step farther, and has pictured the struggle among the determining elements of the germ-cell's organization. It is at least conceivable that the stronger 'determinants,' *i.e.*, the particles embodying the rudiments of certain qualities, will make more of the food-supply than those which are weaker, and that a selective process will ensue" (Thomson). This is the conception of germinal selection.

He has also extended the application of the general doctrine of natural selection by supplying a great number of new illustrations.

The whole theory of Weismann is so well constructed that it is very alluring. Each successive position is worked out with such detail and apt illustration that if one follows him step by step without dissent on some fundamental principle, his conclusion seems justified. As a system it has been elaborated until it makes a coherent appeal to the intellect.

Inheritance of Acquired Characters.—Another fundamental point in Weismann's theory is the denial that acquired characters are transmitted from parent to offspring. Probably the best single discussion of this subject is contained in his book on *The Evolution Theory*, 1904, to which readers are referred.

A few illustrations will be in place. Acquired characters are any acquisitions made by the body-cells during the lifetime of an individual. They may be obvious, as skill in piano-playing, bicycle-riding, etc.; or they may be very recondite, as turns of the intellect, acquired beliefs, etc. Acquired bodily characters may be forcibly impressed upon the organism, as the facial mutilations practiced by certain savage tribes, the docking of the tails of horses, of dogs, etc. The question is, Are any acquired characters, physical or mental, transmitted by inheritance?

Manifestly, it will be difficult to determine on a scientific basis whether or not such qualities are inheritable. One would naturally think first of applying the test of experiment to supposed cases of such inheritances, and this is the best ground to proceed on.

It has been maintained on the basis of the classical experiments of Brown-Séquard on guinea-pigs that induced epilepsy is transmitted to offspring; and, also, on the basis of general observations, that certain bodily mutilations are inherited. Weismann's analysis of the whole situation is very incisive. He experimented by cutting off the tails of both parents of breeding mice. The experiments were carried through twenty-two generations, both parents being deprived of their tails, without yielding any evidence that the mutilations were inheritable.

To take one other case that is less superficial, it is generally believed that the thirst for alcoholic liquors has been transmitted to the children of drunkards, and while Weismann admits the possibility of this, he maintains that it is owing to the germinal elements being exposed to the influence of the alcohol circulating in the blood of the parent or parents; and if this be the case it would not be the inheritance of an acquired character, but the response of the organism to a drug producing directly a variation in the germ-plasm.

Notwithstanding the well-defined opposition of Weismann, the inheritance of acquired characters is still a mooted question. Herbert Spencer argued in favor of it, and during his lifetime had many a pointed controversy with Weismann. Eimer stands unalterably against Weismann's position, and the Neo-Lamarckians stand for the direct inheritance of useful variations in bodily structure. The question is still undetermined and is open to experimental observation. In its present state there are competent observers maintaining both sides, but it must be confessed that there is not a single case in which the supposed inheritance of an acquired character has stood the test of critical examination.

The basis of Weismann's argument is not difficult to understand. Acquired characters affect the body-cells, and according to his view the latter are simply a vehicle for the germinal elements, which are the only things concerned in the transmission of hereditary qualities. Inheritance, therefore, must come through alterations in the germ-plasm, and not directly through changes in the body-cells.

Weismann, the Man.—The man who for more than forty years has been elaborating this theory (Fig. 114) is still living and actively at work in the University of Freiburg. August Weismann was born at Frankfort-on-the-Main in 1834. He was graduated at Göttingen in 1856, and for a short time thereafter engaged in the practice of medicine. This line of activity did not, however, satisfy his nature, and he turned to the pursuit of microscopic investigations in embryology

FIG. 114.—AUGUST WEISMANN, BORN 1834.
Permission of Charles Scribner's Sons.

and morphology, being encouraged in this work by Leuckart, whose name we have already met in this history. In 1863 he settled in Freiburg as *privat-docent,* and has remained connected with the university ever since. From 1867 onward he has occupied the chair of zoölogy in that institution. He has made his department famous, especially by his lectures on the theory of descent.

He is a forceful and interesting lecturer. One of his hearers in 1896 wrote: "His lecture-room is always full, and his popularity among his students fully equals his fame among scientists."

It is quite generally known that Weismann since he reached the age of thirty has been afflicted with an eye-trouble, but the inference sometimes made by those unacquainted with his work as an investigator, that he has been obliged to forego practical work in the field in which he has speculated, is wrong. At intervals his eyes have strengthened so that he has been able to apply himself to microscopic observations, and he has a distinguished record as an observer. In embryology his studies on the development of the diptera, and of the eggs of daphnid crustacea, are well known, as are also his observations on variations in butterflies and other arthropods.

He is an accomplished musician, and during the period of his enforced inactivity in scientific work he found much solace in playing "a good deal of music." "His continuous eye trouble must have been a terrible obstacle, but may have been the prime cause of turning him to the theories with which his name is connected."

In a short autobiography published in *The Lamp* in 1903, although written several years earlier, he gives a glimpse of his family life. "During the ten years (1864–1874) of my enforced inactivity and rest occurred my marriage with Fräulein Marie Gruber, who became the mother of my children and was my true companion for twenty years, until her death. Of her now I think only with love and gratitude. She was the one who, more than any one else, helped me through the gloom of this period. She read much to me at this time, for she read aloud excellently, and she not only took an interest in my theoretical and experimental work, but she also gave practical assistance in it."

In 1893 he published *The Germ-Plasm, A Theory of Heredity*, a treatise which elicited much discussion. From that time on he has been actively engaged in replying to his critics and in perfecting his system of thought.

The Mutation-Theory of De Vries.—Hugo de Vries (Fig. 115), director of the Botanical Garden in Amsterdam, has experimented widely with the growth of plants, especiallv the evening primrose, and has shown that different species appear to rise suddenly. The sudden variations that breed true, and thus give rise to new forms, he calls mutations, and this indicates the source of the name applied to his theory.

In his *Die Mutationstheorie*, published in 1901, he argues for the recognition of mutations as the universal source of the origin of species. Although he evokes natural selection for the perpetuation and improvement of variations, and points out that his theory is not antagonistic to that of natural selection, it is nevertheless directly at variance with Darwin's fundamental conception—that slight individual variations "are probably the sole differences which are effective in the production of new species" and that "as natural selection acts solely by accumulating slight, successive, favorable variations, it can produce no great or sudden modifications." The foundation of De Vries's theory is that "species have not arisen through gradual selection, continued for hundreds or thousands of years, but by jumps through sudden, through small transformations." (Whitman's translation.)

The work of De Vries is a most important contribution to the study of the origin of species, and is indicative of the fact that many factors must be taken into consideration when one attempts to analyze the process of organic evolution. One great value of his work is that it is based on experiments, and that it has given a great stimulus to experimental studies. Experiment was likewise a dominant feature in Darwin's work, but that seems to have been almost overlooked in the discussions aroused by his conclusions; De Vries, by building upon experimental evidence, has led naturalists to

FIG. 115.—HUGO DE VRIES.

realize that the method of evolution is not a subject for argumentative discussion, but for experimental investigation. This is most commendable.

De Vries's theory tends also to widen the field of exploration. Davenport, Tower, and others have made it clear that species may arise by slow accumulations of trivial variations, and that, while the formation of species by mutation may be admitted, there is still abundant evidence of evolution without mutation.

Reconciliation of Different Theories.—All this is leading to a clearer appreciation of the points involved in the discussion of the theories of evolution; the tendency is not for the breach between the different theories to be widened, but for evolutionists to realize more fully the great complexity of the process they are trying to explain, and to see that no single factor can carry the burden of an explanation. Mutation is not a substitute for natural selection, but a coöperating factor; and neither mutation nor natural selection is a substitute for the doctrine of the continuity of the germ-plasm. Thus we may look forward to a reconciliation between apparently conflicting views, when naturalists by sifting shall have determined the truth embodied in the various theories. One conviction that is looming into prominence is that this will be promoted by less argument and more experimental observation.

That the solution of the underlying question in evolution will still require a long time is evident; as Whitman said in his address before the Congress of Arts and Science in St. Louis in 1904: "The problem of problems in biology to-day, the problem which promises to sweep through the present century as it has the past one, with cumulative interest and correspondingly important results, is the one which became the life-work of Charles Darwin, and which can not be better or more simply expressed than in the title of his epoch-making book, *The Origin of Species*."

Summary.—The number of points involved in the four theories considered above is likely to be rather confusing, and we may now bring them into close juxtaposition. The salient features of these theories are as follows:

I. Lamarck's Theory of Evolution.

1. Variation is explained on the principle of use and disuse.
2. Heredity: The variations are inherited directly and improved in succeeding generations.

A long time and favorable conditions are required for the production of new species.

II. Darwin's Theory of Natural Selection.

1. Variations assumed.
2. Heredity: Those slight variations which are of use to the organism will be perpetuated by inheritance.
3. Natural selection is the distinguishing feature of the theory. Through the struggle for existence nature selects those best fitted to survive. The selection of trivial variations that are of advantage to the organism, and their gradual improvement, leads to the production of new species.

III. Weismann's Theory of Continuity of the Germ-plasm.

1. The germ-plasm has had unbroken continuity from the beginning of life. Owing to its impressionable nature, it has an inherited organization of great complexity.
2. Heredity is accounted for on the principle that the offspring is composed of some of the same stuff as its parents. The body-cells are not inherited, *i.e.*,
3. There is no inheritance of acquired characters.
4. Variations arise from the union of the germinal elements, giving rise to varied combinations and permutations of the qualities of the germ-plasm. The purpose of amphimixis is to give rise to variations. The direct influence of environment has produced variations in unicellular organisms.

5. Weismann adopts and extends the principle of natural selection. Germinal selection is exhibited in the germ-plasm.

IV. De Vries's Theory of Mutations.

1. The formation of species is due not to gradual changes, but to sudden mutations.
2. Natural selection presides over and improves variations arising from mutation.

Among the other theories of evolution that of Eimer is the most notable. He maintains that variations in organisms take place not fortuitously or accidentally, but follow a perfectly determinate direction. This definitely directed evolution is called orthogenesis. He insists that there is continuous inheritance of acquired characters, and he is radicallv opposed to the belief that natural selection plays an important part in evolution. The title of his pamphlet published in 1898, *On Orthogenesis and the Impotence of Natural Selection in Species-Formation*, gives an indication of his position in reference to natural selection. A consideration of Eimer's argument would be beyond the purpose of this book.

The cause for the general confusion in the popular mind regarding any distinction between organic evolution and Darwinism is not far to seek. As has been shown, Lamarck launched the doctrine of organic evolution, but his views did not even get a public hearing. Then, after a period of temporary disappearance, the doctrine of evolution emerged again in 1859. And this time the discussion of the general theory centered around Darwin's hypothesis of natural selection. It is quite natural, therefore, that people should think that Darwinism and organic evolution are synonymous terms. The distinction between the general theory and any particular explanation of it has, I trust, been made sufficiently clear in the preceding pages.

CHAPTER XIX

THE RISE OF EVOLUTIONARY THOUGHT

A CURRENT of evolutionary thought can be traced through the literature dealing with organic nature from ancient times. It began as a small rill among the Greek philosophers and dwindles to a mere thread in the Middle Ages, sometimes almost disappearing, but is never completely broken off. Near the close of the eighteenth century it suddenly expands, and becomes a broad and prevailing influence in the nineteenth century. Osborn, in his book, *From the Greeks to Darwin*, traces the continuity of evolutionary thought from the time of the Greek philosophers to Darwin. The ancient phase, although interesting, was vague and general, and may be dismissed without much consideration. After the Renaissance naturalists were occupied with other aspects of nature-study. They were at first attempting to get a knowledge of animals and plants as a whole, and later of their structure, their developments, and their physiology, before questions of their origin were brought under consideration.

Opinion before Lamarck.—The period just prior to Lamarck is of particular interest. Since Lamarck was the first to give a comprehensive and consistent theory of evolution, it will be interesting to determine what was the state of opinion just prior to the appearance of his writings. Studies of nature were in such shape at that time that the question of the origin of species arose, and thereafter it would not recede. This was owing mainly to the fact that Ray and Linnæus by defining a species had fixed the attention of naturalists upon the distinguishing features of the particular kinds of animals and plants. Are species realities in nature? The consideration of this apparently simple question soon led to divergent views, and then to warm controversies that extended over several decades of time.

The view first adopted without much thought and as a matter of course was that species are fixed and constant; *i.e.*, that the existing forms of animals and plants are the descendants of entirely similar parents that were originally created in pairs. This idea of the fixity of species was elevated to the position of a dogma in science as well as in theology. The opposing view, that species are changeable, arose in the minds of a few independent observers and thinkers, and, as has already been pointed out, the discussion of this question resulted ultimately in a complete change of view regarding nature and man's relation to it. When the conception of evolution came upon the scene, it was violently combated. It came into conflict with the theory designated special creation.

Views of Certain Fathers of the Church.—And now it is essential that we should be clear as to the sources of this dogma of special creation. It is perhaps natural to assume that there was a conflict existing between natural science and the views of the theologians from the earliest times; that is, between the scientific method and the method of the theologians, the latter being based on authority, and the former upon observation and experiment. Although there is a conflict between these two methods, there nevertheless was a long period in which many of the leading theological thinkers were in harmony with the men of science with reference to their general conclusions regarding creation. Some of the early Fathers of the Church exhibited a broader and more scientific spirit than their successors.

St. Augustine (353–430), in the fifth century, was the

first of the great theologians to discuss specifically the question of creation. His position is an enlightened one. He says: "It very often happens that there is some question as to the earth or the sky, or the other elements of this world . . . respecting which one who is not a Christian has knowledge derived from most certain reasoning or observation" (that is, a scientific man); "and it is very disgraceful and mischievous and of all things to be carefully avoided, that a Christian speaking of such matters as being according to the Christian Scriptures, should be heard by an unbeliever talking such nonsense that the unbeliever, perceiving him to be as wide from the mark as east from west, can hardly restrain himself from laughing." (Quoted from Osborn.)

Augustine's view of the method of creation was that of derivative creation or creation *causaliter*. His was a natural istic interpretation of the Mosaic record, and a theory of gradual creation. He held that in the beginning the earth and the waters of the earth were endowed with power to produce plants and animals, and that it was not necessary to assume that all creation was formed at once. He cautions his readers against looking to the Scriptures for scientific truths. He said in reference to the creation that the days spoken of in the first chapter of Genesis could not be solar days of twenty-four hours each, but that they must stand for longer periods of time.

This view of St. Augustine is interesting as being less narrow and dogmatic than the position assumed by many theologians of the nineteenth century.

The next theologian to take up the question of creation was St. Thomas Aquinas (1225–1274) in the thirteenth century. He quotes St. Augustine's view with approval, but does not contribute anything of his own. One should not hastily conclude, however, because these views were held by leaders of theological thought, that they were universally accepted. "The truth is that all classes of theologians departed from the original philosophical and scientific stand ards of some of the Fathers of the Church, and that special creation became the universal teaching from the middle of the sixteenth to the middle of the nineteenth centuries."

The Doctrine of Special Creation.—About the seventeenth century a change came about which was largely owing to the writings and influence of a Spanish theologian named Suarez (1548–1617). Although Suarez is not the sole founder of this conception, it is certain, as Huxley has shown, that he engaged himself with the questions raised by the Biblical account of creation; and, furthermore, that he opposed the views that had been expressed by Augustine. In his tract upon the work of the six days (*Tractatus de opere sex dierum*) he takes exception to the views expressed by St. Augustine; he insisted that in the Scriptural account of creation a day of twenty-four hours was meant, and in all other cases he insists upon a literal interpretation of the Scriptures. Thus he introduced into theological thought the doctrine which goes under the name of special creation. The interesting feature in all this is that from the time of St. Augustine, in the fifth century, to the time when the ideas of Suarez began to prevail, in the seventeenth, there had been a harmonious relation between some of the leading theologians and scientific men in their outlook upon creation.

The opinion of Augustine and other theologians was largely owing to the influence of Aristotle. "We know" says Osborn, "that Greek philosophy tinctured early Christian theology; what is not so generally realized is that the Aristotelian notion of the development of life led to the true interpretation of the Mosaic account of the creation.

"There was in fact a long Greek period in the history of the evolutionary idea extending among the Fathers of the Church and later among some of the schoolmen, in their commentaries upon creation, which accord very closely with the modern theistic conception of evolution. If the orthodoxy of Augustine had remained the teaching of the Church, the final establishment of evolution would have come far earlier than it did, certainly during the eighteenth century instead of the nineteenth century, and the bitter controversy over this truth of nature would never have arisen."

The conception of special creation brought into especial prominence upon the Continent by Suarez was taken up by John Milton in his great epic *Paradise Lost*, in which he gave a picture of creation that molded into specific form the opinion of the English-speaking clergy and of the masses who read his book. When the doctrine of organic evolution was announced, it came into conflict with this particular idea; and, as Huxley has very pointedly remarked, the new theory of organic evolution found itself in conflict with the Miltonic, rather than the Mosaic cosmology. All this represents an interesting phase in intellectual development.

Forerunners of Lamarck.—We now take up the immediate predecessors of Lamarck. Those to be mentioned are Buffon, Erasmus Darwin, and Goethe.

Buffon (1707–1788) (Fig. 116), although of a more philosophical mind than many of his contemporaries, was not a true investigator. That is, he left no technical papers or contributions to science. From 1739 to the time of his death he was the superintendent of the *Jardin du Roi*. He was a man of elegance, with an assured position in society. He was a delightful writer, a circumstance that enabled him to make natural history popular. It is said that the advance sheets of Buffon's *Histoire Naturelle* were to be found on the tables of the boudoirs of ladies of fashion. In that work he suggested the idea that the different forms of life were gradually produced, but his timidity and his prudence led him to be obscure in what he said.

Packard, who has studied his writings with care, says that he was an evolutionist through all periods of his life, not, as is commonly maintained, believing first in the fixitv of species, later in their changeability, and lastly returning to his earlier position. "The impression left on the mind after

FIG. 116.—BUFFON, 1707–1788.

reading Buffon is that even if he threw out these suggestions and then retracted them, from fear of annoyance or even persecution from the bigots of his time, he did not himself always take them seriously, but rather jotted them down as passing thoughts. Certainlv he did not present them in the formal, forcible, and scientific way that Erasmus Darwin did. The result is that the tentative views of Buffon, which have to be with much research extracted from the forty-four volumes of his works, would now be regarded as in a degree superficial and valueless. But they appeared thirty-four years before Lamarck's theory, and though not epoch-making,

FIG. 117.—ERASMUS DARWIN, 1731–1802.

they are such as will render the name of Buffon memorable for all time" (Packard.)

Erasmus Darwin (Fig. 117) was the greatest of Lamarck's predecessors. In 1794 he published the *Zoönomia*. In this work he stated ten principles; among them he vaguely suggested the transmission of acquired characteristics, the law of sexual selection—or the law of battle, as he called it—

protective coloration, etc. His work received some notice from scholars. Paley's *Natural Theology*, for illustration, was written against it, although Paley is careful not to mention Darwin or his work. The success of Paley's book is probably one of the chief causes for the neglect into which the views of Buffon and Erasmus Darwin fell.

Inasmuch as Darwin's conclusions were published before Lamarck's book, it would be interesting to determine whether or not Lamarck was influenced by him. The careful consideration of this matter leads to the conclusion that Lamarck drew his inspiration directly from nature, and that points of similarity between his views and those of Erasmus Darwin are to be looked upon as an example of parallelism in thought. It is altogether likely that Lamarck was wholly unacquainted with Darwin's work, which had been published in England.

Goethe's connection with the rise of evolutionary thought is in a measure incidental. In 1790 he published his *Metamorphosis of Plants*, showing that flowers are modified leaves. This doctrine of metamorphosis of parts he presently applied to the animal kingdom, and brought forward his famous, but erroneous, vertebrate theory of the skull. As he meditated on the extent of modifications there arose in his mind the conviction that all plants and animals have been evolved from the modification of a few parental types. Accordingly he should be accorded a place in the history of evolutionary thought.

Opposition to Lamarck's Views.—Lamarck's doctrine which was published in definite form in 1809, has been already outlined. We may well inquire, Why did not his views take hold? In the first place, they were not accepted by Cuvier. Cuvier's opposition was strong and vigorous. and succeeded in causing the theory of Lamarck to be completely neglected by the French people. Again, we must recognize that the time was not ripe for the acceptance of such truths; and, finally, that there was no great principle enunciated by Lamarck which could be readily understood as there was in Darwin's book on the doctrine of natural selection.

The temporary disappearance of the doctrine of organic evolution which occurred after Lamarck expounded his theory was also owing to the reaction against the speculations of the school of *Natur-Philosophie*. The extravagant speculation of Oken and the other representatives of this school completely disgusted men who were engaged in research by observation and experiment. The reaction against that school was so strong that it was difficult to get a hearing for any theoretical speculation; but Cuvier's influence must be looked upon as the chief one in causing disregard for Lamarck's writings.

The work of Cuvier has been already considered in connection both with comparative anatomy and zoölogy, but a few points must still be held under consideration. Cuvier brought forward the idea of catastrophism in order to explain the disappearance of the groups of fossil animals. He believed in the doctrine of spontaneous generation. He held to the doctrine of pre-delineation, so that it must be admitted that whenever he forsook observation for speculation he was singularly unhappy; and it is undeniable that his position of hostility in reference to the speculation of Lamarck retarded the progress of science for nearly half a century.

Cuvier and Saint-Hilaire.—In 1830 there occurred a memorable controversy between Cuvier and Saint-Hilaire. The latter (Fig. 118) was in early life closely associated with Lamarck, and shared his views in reference to the origin of animals and plants; though in certain points Saint-Hilaire was more a follower of Buffon than of Lamarck. Strangely enough, Saint-Hilaire was regarded as the stronger man of the two. He was more in the public eye, but was not a man of such deep intellectuality as Lamarck. His scientific reputation rests mainly upon his *Philosophie Anatomique*. The controversy between him and Cuvier was on the subject of

FIG. 118.—GEOFFROY SAINT-HILAIRE, 1772–1844.

unity of type; but it involved the question of the fixity or mutability of species, and therefore it involved the foundation of the question of organic evolution.

This debate stirred all intellectual Europe. Cuvier won as being the better debater and the better manager of his case. He pointed triumphantly to the four branches of the animal kingdom which he had established, maintaining that these four branches represented four distinct types of organization; and, furthermore, that fixity of species and fixity of type were necessary for the existence of a scientific natural history. We can see now that his contention was wrong, but at the time he won the debate. The young men of the period, that is, the rising biologists of France, were nearly all adherents of Cuvier, so that the effect of the debate was, as previously stated, to retard the progress of science. This noteworthy debate occurred in February, 1830. The wide and lively interest with which the debate was followed may be inferred from the excitement manifested by Goethe. Of the great poet-naturalist, who was then in his eighty-first year, the following incident is told by Soret:

"Monday, Aug. 2d, 1830.—The news of the outbreak of the revolution of July arrived in Weimar to-day, and has caused general excitement. In the course of the afternoon I went to Goethe. 'Well,' he exclaimed as I entered, 'what do you think of this great event? The volcano has burst forth, all is in flames, and there are no more negotiations behind closed doors.' 'A dreadful affair,' I answered; 'but what else could be expected under the circumstances, and with such a ministry, except that it would end in the expulsion of the present royal family?' 'We do not seem to understand each other, my dear friend,' replied Goethe. 'I am not speaking of those people at all; I am interested in something very different. I mean the dispute between Cuvier and Geoffroy de Saint-Hilaire, which has broken out in the Academy, and which is of such great importance to science.' This remark of Goethe came upon me so unexpectedly that I did not know what to say, and my thoughts for some minutes seemed to have come to a complete standstill. 'The affair is of the utmost importance,' he continued, 'and you can not form any idea of what I felt on receiving the news of the meeting on the 19th. In Geoffroy de Saint-Hilaire we have now a mighty ally for a long time to come. But I see also how great the sympathy of the French scientific world must be in this affair, for, in spite of the terrible political excitement, the meeting on the 19th was attended by a full house. The best of it is, however, that the synthetic treatment of nature, introduced into France by Geoffroy, can now no longer be stopped. This matter has now become public through the discussions in the Academy, carried on in the presence of a large audience; it can no longer be referred to secret committees, or be settled or suppressed behind closed doors.' "

Influence of Lyell's Principles of Geology.—But just as Cuvier was triumphing over Saint-Hilaire a work was being published in England which was destined to overthrow the position of Cuvier and to bring again a sufficient foundation for the basis of mutability of species. I refer to Lyell's *Principles of Geology*, the influence of which has already been spoken of in Chapter XV. Lyell laid down the principle that we are to interpret occurrences in the past in the terms of what is occurring in the present. He demonstrated that observations upon the present show that the surface of the earth is undergoing gradually slow changes through the action of various agents, and he pointed out that we must view the occurrences in the past in the light of occurrences in the present. Once this was applied to animal forms it became evident that the observations upon animals and plants in the present must be applied to the life of the fossil series.

These ideas, then, paved the way for the conception of changes in nature as being one continuous series.

H. Spencer.—In 1852 came the publication of Herbert Spencer in the *Leader*, in which he came very near anticipating the doctrine of natural selection. He advanced the developmental hypothesis, saying that even if its supporters could "merely show that the production of species by the process of modification is conceivable, they would be in a better position than their opponents. But they can do much more than this; they can show that the process of modification has affected and is affecting great changes in all organisms subject to modifying influences. . . . They can show that any existing species, animal or vegetable, when placed under conditions different from its previous ones, immediately begins to undergo certain changes of structure fitting it for the new conditions. They can show that in successive generations these changes continue, until ultimately the new conditions become the natural ones. They can show that in cultivated plants and domesticated animals, and in the several races of men, these changes have uniformly taken place. They can show that the degrees of difference so produced are often, as in dogs, greater than those on which distinctions of species are in other cases founded. They can show that it is a matter of dispute whether some of these modified forms *are* varieties or modified species. And thus they can show that throughout all organic nature there is at work a modifying influence of the kind they assign as the cause of these specific differences; an influence which, though slow in its action, does in time, if the circumstances demand it, produce marked changes; an influence which, to all appearance, would produce in the millions of years, and under the great varieties of conditions which geological records imply, any amount of change."

'It is impossible," says Marshall, "to depict better than this the condition prior to Darwin. In this essay there is full recognition of the fact of transition, and of its being due to natural influences or causes, acting now and at all times. Yet it remained comparatively unnoticed, because Spencer, like his contemporaries and predecessors, while advocating evolution, was unable to state explicitly what these causes were."

Darwin and Wallace.—In 1858 we come to the crown ing event in the rise of evolutionary thought, when Alfred Russel Wallace sent a communication to Mr. Darwin, begging him to look it over and give him his opinion of it. Darwin, who had been working upon his theory for more than twenty years, patiently gathering facts and testing the same by experiment, was greatly surprised to find that Mr. Wallace had independently hit upon the same principle of explaining the formation of species. In his generosity, he was at first disposed to withdraw from the field and publish the essay of Wallace without saying anything about his own work. He decided, however, to abide by the decision of two of his friends, to whom he had submitted the matter, and the result was that the paper of Wallace, accompanied by earlier communications of Darwin, were laid before the Linnæan Society of London. This was such an important event in the history of science that its consideration is extended by quoting the following letter:

"LONDON, June 30th, 1858.

"MY DEAR SIR: The accompanying papers, which we have the honor of communicating to the Linnæan Society, and which all relate to the same subject; *viz.*, the laws which affect the production of varieties, races, and species, contain the results of the investigations of two indefatigable naturalists, Mr. Charles Darwin and Mr. Alfred Wallace.

"These gentlemen having, independently and unknown to one another, conceived the same very ingenious theory to account for the appearance and perpetuation of varieties and of specific forms on our planet, may both fairly claim the merit of being original thinkers in this important line of inquiry; but neither of them having published his views, though Mr. Darwin has for many years past been repeatedly urged by us to do so, and both authors having now unreservedly placed their papers in our hands, we think it would best promote the interests of science that a selection from them should be laid before the Linnæan Society.

"Taken in the order of their dates, they consist of:

"1. Extracts from a MS. work on species, by Mr. Darwin, which was sketched in 1839 and copied in 1844, when the copy was read by Dr. Hooker, and its contents afterward communicated to Sir Charles Lyell. The first part is devoted to *The Variation of Organic Beings under Domestication and in their Natural State;* and the second chapter of that part, from which we propose to read to the Society the extracts referred to, is headed *On the Variation of Organic Beings in a State of Nature; on the Natural Means of Selection; on the Comparison of Domestic Races and True Species.*

"2. An abstract of a private letter addressed to Professor Asa Gray, of Boston, U. S., in October, 1857, by Mr. Darwin, in which he repeats his views, and which shows that these remained unaltered from 1839 to 1857.

"3. An essay by Mr. Wallace, entitled *On the Tendency of Varieties to Depart Indefinitely from the Original Type.* This was written at Ternate in February, 1858, for the perusal of his friend and correspondent, Mr. Darwin, and sent to him with the expressed wish that it should be forwarded to Sir Charles Lyell, if Mr. Darwin thought it sufficiently novel and interesting. So highly did Mr. Darwin appreciate the value of the views therein set forth that he proposed, in a letter to Sir Charles Lyell, to obtain Mr. Wallace's consent to allow the essay to be published as soon as possible. Of this step we highly approved, provided Mr. Darwin did not withhold from the public, as he was strongly inclined to do (in favor of Mr. Wallace), the memoir which he had himself written on the same subject, and which, as before stated, one of us had perused in 1844, and the contents of which we had both of us been privy to for many years.

"On representing this to Mr. Darwin, he gave us permission to make what use we thought proper of his memoir, etc.; and in adopting our present course, of presenting it to the Linnæan Society, we have explained to him that we are not solely considering the relative claims to priority of himself and his friend, but the interests of science generally; for we feel it to be desirable that views founded on a wide deduction from facts, and matured by years of reflecting, should constitute at once a goal from which others may start; and that, while the scientific world is waiting for the appearance of Mr. Darwin's complete work, some of the leading results of his labours, as well as those of his able correspondent, should together be laid before the public.

"We have the honour to be yours very obediently,

CHARLES LYELL
JOS. D. HOOKER."

Personality of Darwin.—The personality of Darwin is extremely interesting. Of his numerous portraits, the one shown in Fig. 119 is less commonly known than those showing him with a beard and a much furrowed forehead. This portrait represents him in middle life, about the time of the publication of his *Origin of Species.* It shows a rather typical British face, of marked individuality. Steadiness, sincerity, and urbanity are all depicted here. His bluish-gray eyes were overshadowed by a projecting ridge and very prominent, bushy eyebrows that make his portrait, once seen, easily recognized thereafter. In the full-length portraits representing him seated, every line in his body shows the quiet, philosophical temper for which he was notable. An intimate account of his life is contained in the *Life and Letters of Charles Darwin* (1887) and in *More Letters of Darwin* (1903), both of which are illustrated by portraits and other pictures. The books about Darwin and his work are numerous, but the reader is referred in particular to the two mentioned as giving the best conception of the great naturalist and of his personal characteristics.

He is described as being about six feet high, but with a stoop of the shoulders which diminished his apparent height;

FIG. 119—CHARLES DARWIN, 1809–1882.

"of active habits, but with no natural grace or neatness of movement." "In manner he was bright, animated, and cheerful; a delightfully considerate host, a man of never-failing courtesy, leading him to reply at length to letters from anybody, and sometimes of a most foolish kind."

His Home Life.—"Darwin was a man greatly loved and respected by all who knew him. There was a peculiar charm about his manner, a constant deference to others, and a faculty for seeing the best side of everything and everybody."

He was most affectionate and considerate at home. The picture of Darwin's life with his children gives a glimpse of the tenderness and deep affection of his nature, and the reverent regard with which he was held in the family circle is very touching. One of his daughters writes: "My first remembrances of my father are of the delights of his playing with us. He was passionately attached to his own children, although he was not an indiscriminate child-lover. To all of us he was the most delightful playfellow, and the most perfect sympathizer. Indeed, it is impossible adequately to describe how delightful a relation his was to his family, whether as children or in their later life.

"It is a proof of the terms on which we were, and also of how much he was valued as a playfellow, that one of his sons, when about four years old, tried to bribe him with a sixpence to come and play in working hours. We all knew the sacredness of working time, but that any one should resist sixpence seemed an impossibility."

Method of Work.—Darwin's life, as might be inferred from the enduring quality of his researches, shows an unswerving purpose. His theory was not the result of a sudden flash of insight, nor was it struck out in the heat of inspiration, but was the product of almost unexampled industry and conscientious endeavor in the face of unfavorable circumstances. Although strikingly original and independent as a thinker, he was slow to arrive at conclusions, examining with the most minute and scrupulous care the ground for every conclusion. "One quality of mind that seemed to be of especial advantage in leading him to make discoveries was the habit of never letting exceptions pass unnoticed." He enjoyed experimenting much more than work which only entailed reasoning. Of course, he was a great reader, but for books as books he had no respect, often cutting large ones in two in order to make them easier to hold while in use.

Darwin's Early Life.—Charles Darwin was born in 1809 at Shrewsbury, England, of distinguished ancestry, his grandfather being the famous Dr. Erasmus Darwin, the founder, as we have seen, of a theory of evolution. In his youth he gave no indication of future greatness. He was sent to Edinburgh to study medicine, but left there after two sessions, at the suggestion of his father, to study for the Church. He then went to the University of Cambridge, where he remained three years, listening to "incredibly dull lectures." After taking his baccalaureate degree, came the event which proved, as Darwin says, "the turning-point of my life." This was his appointment as naturalist on the surveying expedition about to be entered upon by the ship *Beagle*. In Cambridge he had manifested an interest in scientific study, and had been encouraged by Professor Henslow, to whom he was also indebted for the recommendation to the post on the *Beagle*. An amusing circumstance connected with his appointment is that he was nearly rejected by Captain Fitz-Rov, who doubted "whether a man with such a shaped nose could possess sufficient energy and determination for the voyage."

Voyage of the Beagle.—The voyage of the *Beagle* extended over five years (1831–1836), mainly along the west coast of South America. It was on this voyage that Darwin acquired the habit of constant industrv. He had also opportunity to take long trips on shore, engaged in observation and in making extensive collections. He observed nature in the field under exceptional circumstances. As he traveled he noted fossil forms in rocks as well as the living forms in field and forest. He observed the correspondence in type between certain extinct forms and recent animals in South America. He noticed in the Galapagos Islands a fauna similar in general characteristics to that of the mainland, five or six hundred miles distant, and yet totally different as to species. Moreover, certain species were found to be confined to particular islands. These observations awakened in his mind, a mind naturally given to inquiring into the causes of things, questions that led to the formulation of his theory. It was not, however, until 1837 that he commenced his first note book for containing his observations upon the transmutations of animals. He started as a firm believer in the fixity of species, and spent several years collecting and considering data before he changed his views.

At Downs.—On his return to England, after spending some time in London, he purchased a country-place at Downs, and, as his inheritance made it possible, he devoted himself entirely to his researches.

But, as is well known, he found in his illness a great obstacle to steady work. He had been a vigorous youth and young man, fond of outdoor sports, as fishing, shooting, and the like. After returning from his long voyage, he was affected by a form of constant illness, involving a giddiness in the head, and "for nearly forty years he never knew one day of the health of an ordinary man, and thus his life was one long struggle against the weariness and strain of sickness." Gould in his *Biographical Clinics* attributes his illness to eye-strain.

"Under such conditions absolute regularity of routine was essential, and the day's work was carefully planned out. At his best, he had three periods of work: from 8.00 to 9.30; from 10.30 to 12.15; and from 4.30 to 6.00, each period being under two hours' duration."

The patient thoroughness of his experimental work and of his observation is shown by the fact that he did not publish his book on the *Origin of Species* until he had worked on his theory twenty-two years. The circumstances that led to his publishing it when he did have already been indicated.

Parallelism in the Thought of Darwin and Wallace.—No one can read the letters of Darwin and Wallace explaining how they arrived at their idea of natural selection without marveling at the remarkable parallelism in the thought of the two. It is a noteworthy circumstance that the idea of natural selection came to both by the reading of the same book, *Malthus on Population.*

Darwin's statement of how he arrived at the conception of natural selection is as follows: "In October, 1838, that is, fifteen months after I had begun my systematic inquiry, I happened to read for amusement *Malthus on Population*, and being well prepared to appreciate the struggle for existence which everywhere goes on from long-continued observations of the habits of animals and plants, it at once struck me that under these circumstances favourable variations would tend to be preserved and unfavourable ones to be destroyed. *The result of this would be the formation of new species.* Here then I had at last got a theory by which to work, but I was so anxious to avoid prejudice that I determined not for some time to write even the briefest sketch of it. In June, 1842, I first allowed myself the satisfaction of writing a very brief abstract of my theory in pencil, in thirty-five pages, and this was enlarged during the summer of 1844 into one of 230 pages."

And Wallace gives this account: "In February, 1858, I was suffering from a rather severe attack of intermittent fever at Ternate, in the Moluccas; and one day, while lying on my bed during the cold fit, wrapped in blankets, though the thermometer was at 88° Fahr., the problem again presented itself to me, and something led me to think of the 'positive checks' described by Malthus in his *Essay on Population*, a work I had read several years before, and which had made a deep and permanent impression on my mind. These checks—war, disease, famine, and the like—must, it occurred to me, act on animals as well as man. Then I thought of

FIG. 120.—ALFRED RUSSEL WALLACE, BORN 1823.

the enormously rapid multiplication of animals, causing these checks to be much more effective in them than in the case of man; and while pondering vaguely on this fact, there suddenly flashed upon me the *idea* of the survival of the fittest—that the individuals removed by these checks must be on the whole inferior to those that survived. In the two hours that elapsed before my ague fit was over, I had thought out almost the whole of the theory; and the same evening I sketched the draught of my paper, and in the two succeeding evenings wrote it out in full, and sent it by the next post to Mr. Darwin."

It thus appears that the announcement of the Darwin-Wallace theory of natural selection was made in 1858, and in the following year was published the book, the famous *Origin of Species,* upon which Darwin had been working when he received Mr. Wallace's essay. Darwin spoke of this work as an outline, a sort of introduction to other works that were in the course of preparation. His subsequent works upon *Animals and Plants under Domestication, The Descent of Man,* etc., etc., expanded his theory, but none of them effected so much stir in the intellectual world as the *Origin of Species.*

This skeleton outline should be filled out by reading *Darwin's Life and Letters,* by his son, and the complete papers of Darwin and Wallace, as originally published in the *Journal of the Linnæan Society.* The original papers are reproduced in the *Popular Science Monthly* for November, 1901.

Wallace was born in 1823, and is still living. He shares with Darwin the credit of propounding the theory of natural selection, and he is notable also for the publication of important books, as the *Malay Archipelago, The Geographical Distribution of Animals, The Wonderful Century,* etc.

The Spread of the Doctrine of Organic Evolution. Huxley.—Darwin was of a quiet habit, not aggressive in the defense of his views. His theory provoked so much opposition that it needed some defenders of the pugnacious type. In England such a man was found in Thomas Henry Huxley (1825–1895). He was one of the greatest popular exponents of science of the nineteenth century; a man of most thorough and exact scholarship, with a keen, analvtical mind that went directlv to the center of questions under consideration, and powers as a writer that gave him a wide circle of readers. He was magnificently sincere in his fight for the prevalence

FIG. 121.—THOMAS HENRY HUXLEY, 1825–1895.

of intellectual honesty. Doubtless he will be longer remembered for this service than for anything else.

He defended the doctrine of evolution, not only against oratorical attacks like that of Bishop Wilberforce, but against well-considered arguments and more worthy opponents. He advanced the standing of the theory in a less direct way by urging the pursuit of scientific studies by high-school and university students, and by bringing science closer to the people. He was a pioneer in the laboratory teaching of biology, and his *Manual* has been, ever since its publication in 1874, the inspiration and the model for writers of directions for practical work in that field.

It is not so generallv known that he was also a great investigator, producing a large amount of purely technical researches. After his death a memorial edition of his seientific memoirs was published in four large quarto volumes. The extent of his scientific output when thus assembled was a surprise to manv of his co-workers in the field of science. His other writings of a more general character have been collected in fourteen quarto volumes. Some of the essays in this collection are models of clear and vigorous English style. Mr. Huxlev did an astonishing amount of scientific work, especially in morphology and palæontology. Those who have been privileged to look over his manuscripts and unpublished drawings in his old room at South Kensington could not fail to have been impressed, not only with the extent, but also with the accuracy of his work. Taking Johannes Müller as his exemplar, he investigated animal organisms with a completeness and an exactness that have rarely been equaled.

An intimate account of his life will be found in *The Life and Letters of Thomas Henry Huxley*, by his son.

Haeckel.—Ernst Haeckel, of Jena, born in 1834 (Fig. 122), was one of the earliest in Germany to take up the defense of Darwin's hypothesis. As early as 1866 he applied the doctrine of evolution to all organisms in his *Generelle Morphologie.* This work, which has been long out of print, represents his best contribution to evolutionary thought. He has written widely for general readers, and although his writings are popularly believed to represent the best scientific thought on the matter, those written for the general public are not regarded by most biologists as strictly representative.

As a thinker he is more careless than Huxley, and as a result less critical and exact as a writer.

There can be no doubt that the germs of evolutionary thought existed in Greek philosophy, and that they were

FIG. 122.—ERNST HAECKEL, BORN 1834.

retained in a state of low vitality among the mediæval thinkers who reflected upon the problem of creation. It was not, however, until the beginning of the nineteenth century that, under the nurture of Lamarck, they grew into what we may speak of as the modern theory of evolution. After various vicissitudes this doctrine was made fertile by Darwin, who supplied it with a new principle, that of natural selection.

The fruits of this long growth are now being gathered. After Darwin the problem of biology became not merely to describe phenomena, but to explain them. This is the outcome of the rise and progress of biology: first, crude and uncritical observations of the forms of animated nature; then descriptive analysis of their structure and development; and, finally, experimental studies, the effort to explain vital phenomena, an effort in which biologists are at present engaged.

CHAPTER XX

RETROSPECT AND PROSPECT. RECENT TENDENCIES IN BIOLOGY

When one views the progress of biology in retrospect, the broad truth stands out that there has been a continuity of development in biological thought and interpretation. The new proceeds out of the old, but is genetically related to it. A good illustration of this is seen in the modified sense in which the theories of epigenesis and pre-formation have been retained in the biological philosophy of the nineteenth century. The same kind of question that divided the philosophers of the seventeenth and eighteenth centuries has remained to vex those of the nineteenth; and, although both processes have assumed a different aspect in the light of germinal continuity, the theorists of the last part of the nineteenth century were divided in their outlook upon biological processes into those of the epigenetic school and those who are persuaded of a pre-organization in the germinal elements of organisms. Leading biological questions were warmly discussed from these different points of view.

In its general character the progress of natural science has been, and still is, a crusade against superstition; and it may be remarked in passing that "the nature of superstition consists in a gross misunderstanding of the causes of natural phenomena." The struggle has been more marked in biology than in other departments of science because biology involves the consideration of living organisms and undertakes to establish the same basis for thinking about the organization of the human body as about the rest of the animal series.

The first triumph of the scientific method was the overthrow of authority as a means of ascertaining truth and sub stituting therefor the method of observation and experiment. This carries us back to the days of Vesalius and Harvey, before the framework of biology was reared. But the seientific method, once established, led on graduallv to a belief in the constancy of nature and in the prevalence of universal laws in the production of all phenomena. In its progress biology has exhibited three phases which more or less overlap: The first was the descriptive phase, in which the obvious features of animals and plants were merelv described; the descriptive was supplemented bv the comparative method; this in due course bv the experimental method, or the study of the processes that take place in organisms. Thus, description, comparison, and experiment represent the great phases of biological development.

The Notable Books of Biology and their Authors.--The progress of biology has been owing to the efforts of men of very human qualities, yet each with some special distinguishing feature of eminence. Certain of their publications are the mile-stones of the way. It may be worth while, therefore, in a brief recapitulation to name the books of widest general influence in the progress of biology. Onlv those publications will be mentioned that have formed the starting-point of some new movement, or have laid the foundation of some new theory.

Beginning with the revival of learning, the books of Vesalius, *De Corporis Humani Fabrica* (1543), and Harvey, *De Motu Cordis et Sanguinis* (1628), laid the foundations of scientific method in biology.

The pioneer researches of Malpighi on the minute anatomy of plants and animals, and on the development of the chick, best represent the progress of investigation between Harvey and Linnæus. The three contributions referred to are those on the *Anatomy of Plants* (*Anatome Plantarum*, 1675–1679); on the *Anatomy of the Silkworm* (*De Bombyce*, 1669); and on the *Development of the Chick* (*De Formatione Pulli in Ovo* and *De Ovo Incubato*, both 1672).

We then pass to the *Systema Naturæ* (twelve editions, 1735–1768) of Linnæus, a work that had such wide influence in stimulating activity in systematic botany and zoölogy.

Wolff's *Theoria Generationis*, 1759, and his *De Formatione Intestinorum*, 1764, especially the latter, were pieces of observation marking the highest level of investigation of development prior to that of Pander and Von Baer.

Cuvier, in *Le Règne Animal*, 1816, applied the principles of comparative anatomy to the entire animal kingdom.

The publication in 1800 of Bichat's *Traité des Membranes* created a new department of anatomy, called histology.

Lamarck's book, *La Philosophie Zoologique*, 1809, must have a place among the great works in biology. Its influence was delayed for more than fifty years after its publication.

The monumental work of Von Baer on *Development* (*Ueber Entwicklungsgeschichte der Thiere*), 1828, is an almost ideal combination of observation and conclusion in embryology.

The *Microscopische Untersuchungen*, 1839, of Schwann marks the foundation of the cell-theory.

The *Handbook* of Johannes Müller (*Handbuch der Physiologie des Menschen*), 1846, remains unsurpassed as to its plan and its execution.

Max Schultze in his treatise *Ueber Muskelkörperchen und das was man eine Zelle zu nennen habe*, 1861, established one of the most important conceptions with which biology has been enriched, *viz.*, the protoplasm doctrine.

Darwin's *Origin of Species*, 1859, is, from our present outlook, the greatest classic in biology.

Pasteur's *Studies on Fermentation*, 1876, is typical of the quality of his work, though his later investigations on inoculations for the prevention of hydrophobia and other maladies are of greater importance to mankind.

It is somewhat puzzling to select a man to represent the study of fossil life, one is tempted to name E. D. Cope, whose researches were conceived on the highest plane. Zittel, however, covered the entire field of fossil life, and his *Handbook of Palæontology* is designated as a mile-post in the development of that science.

Before the Renaissance the works of Aristotle and Galen should be included.

From the view-point suggested, the more notable figures in the development of biology are: Aristotle, Galen, Vesalius, Harvey, Malpighi, Linnæus, Wolff, Cuvier, Bichat, Lamarck, Von Baer, J. Müller, Schwann, Schultze, Darwin, Pasteur, and Cope.

Such a list is, as a matter of course, arbitrary, and can serve no useful purpose except that of bringing into combination in a single group the names of the most illustrious founders of biological science. The individuals mentioned are not all of the same relative rank, and the list should be extended rather than contracted. Schwann, when the entire output of the two is considered, would rank lower as a scientific man than Koelliker, who is not mentioned, but the former must stand in the list on account of his connection with the cell-theory. Virchow, the presumptive founder of pathology, is omitted, as are also investigators like Koch, whose line of activity has been chiefly medical.

Recent Tendencies in Biology. Higher Standards.—In attempting to indicate some of the more evident influences that dominate biological investigation at the present time, nothing more than an enumeration of tendencies with a running commentary is possible. One notes first a wholesome influence in the establishment of higher standards, both of research and of scientific publication. Investigations as a whole have become more intensive and more critical. Much of the work that would have passed muster for publication two decades ago is now regarded by the editors of the best biological periodicals as too general and too superficial. The requisites for the recognition of creditable work being higher, tends to elevate the whole level of biological science.

Improvement in Tools and Methods.—This has come about partly through improvement in the tools and in the methods of the investigators. It can hardly be said, however, that thinking and discernment have been advanced at the same rate as the mechanical helps to research. In becoming more intensive, the investigation of biological problems has lost something in comprehensiveness. That which some of the earlier investigators lacked in technique was compensated for in the breadth of their preliminary training and in their splendid appreciation of the relations of the facts at their disposal.

The great improvement in the mechanical adjustments and in the optical powers of microscopes has made it possible to see more regarding the physical structure and the activities of organisms than ever before. Microtomes of the best workmanship have placed in the hands of histologists the means of making serial sections of remarkable thinness and regularity.

The great development of micro-chemical technique also has had the widest influence in promoting exact researches in biology. Special staining methods, as those of Golgi and Bethe, by means of which the wonderful fabric of the nervous system has been revealed, are illustrations.

The separation by maceration and smear preparation of entire histological elements so that they may be viewed as solids has come to supplement the study of sections. Reconstruction, by carving wax plates of known thickness into the form of magnified sections drawn upon their surfaces to a scale, and then fitting the plates together, has been very helpful in picturing complicated anatomical relations. This method has made it possible to produce permanent wax models of minute structures magnified to any desired degree. Minute dissections, although not yet sufficiently practiced, are nevertheless better than the wax models for making accurate drawings of minute structures as seen in relief.

The injection of the blood-vessels of extremely small embryos has made it possible to study advantageously the circulatory system. The softening of bones by acid after the tissues are already embedded in celloidin has offered a means of investigating the structure of the internal ear by sections, and is widely applicable to other tissues.

With the advantage of the new appliances and the new methods, the old problems of anatomy are being worked over on a higher level of requirement. Still, it is doubtful whether even the old problems will be solved in more than a relative way. It is characteristic of the progress of research that as one proceeds the horizon broadens and new questions spring up in the pathway of the investigator. He does not solve the problems he sets out to solve, but opens a lot of new ones. This is one of the features of scientific research that make its votaries characteristically optimistic.

Experimental Work.—Among the recent influences tending to advance biology, none is more important than the application of experiments to biological studies. The experimental method is in reality applicable to diverse fields of biological research, and its extensive use at present indicates a movement in the right direction; that is, a growing interest in the study of processes. One of the earliest problems of the biologist is to investigate the architecture of living beings; then there arise questions as to the processes that occur within the organism, and the study of processes involves the employment of experiments. In the pursuit of physiology experiments have been in use since the time of Harvey, but even in that science, where they are indispensable, experiments did not become comparative until the nineteenth century. It now appears that various forms of experiment give also a better insight into the structure of organisms, and the practice of applying experiments to structural studies has given rise to the new department of experimental morphology.

For the purpose of indicating some of the directions in which biology has been furthered by the experimental method of investigation, we designate the fields of heredity and evolution, changes in the environment of organisms, studies on fertilization and on animal behavior.

The recognition that both heredity and the process of evolution can be subjected to experimental tests was a revelation. Darwin and the early evolutionists thought the evolutionary changes too slow to be appreciated, but now we know that many of the changes can be investigated by experiment. Numerous experiments on heredity in poultry (Davenport), in rats, in rabbits, and in guinea-pigs (Castle) have been carried out—experiments that test the laws of ancestral inheritance and throw great light upon the questions introduced by the investigations of Mendel and De Vries. The investigations of De Vries on the evolution of plant-life occupy a notable position among the experimental studies.

A large number of experiments on the effects produced by changes in the external conditions of life have been made. To this class of investigations belong studies on the regulation of form and function in organisms (Loeb, Child), the effects produced by altering mechanical conditions of growth, by changing the chemical environment, etc. There is some internal mechanism in living matter that is influenced by changes in external conditions, and the study of the regulation cf the internal processes that produce form and structure have given rise to a variety of interesting problems. The regeneration of lost parts and regeneration after intentionally-imposed injury has received much attention (Morgan). Marine animals are especially amenable to manipulations of this nature, as well as to alterations in their surroundings, on account of the ease in altering the chemical environment in which they live. The latter may be accomplished by dissolving harmless chemical salts in the sea-water, and observing the changes produced by the alterations of the surrounding conditions. By this means Herbst and others have produced very interesting results.

In the field of artificial fertilization, free swimming larvæ have been raised from eggs artificially fertilized by changes in osmotic pressure, and also by treating them with both organic and inorganic acids; and these studies have greatly altered opinion regarding the nature of fertilization, and of certain other phenomena of development.

Animal Behavior.—The study of animal behavior (Jennings) is a very characteristic activity of the present, in which certain psychological processes are investigated. These investigations have given rise to a distinct line of research participated in by psychologists and biologists. The study of the way in which animals will react toward light of different colors, to variations in the intensity of light, to alterations in temperature, and to various other forms of stimuli are yielding very important results, that enable investigators to look beneath the surface and to make important deductions regarding the nature of psychological processes.

A line closely allied to experimentation is the application of statistics to biological processes, such as those of growth, stature, the law of ancestral inheritance, the statistical study of variations in spines, markings on shells, etc., etc. (Galton, Pearson, Davenport).

Other branches of biology that have been greatly developed by the experimental method are those of bacteriology and physiological chemistry. The advances in the latter have greatly widened the horizon of our view regarding the nature of vital activities, and they compose one of the leading features of current biological investigation.

Some Tendencies in Anatomical Studies. Cell-Lineage.— While experimental work occupies the center of the stage, at the same time great improvements in morphological studies are evident. It will be only possible, however, to indicate in a general way the direction in which investigations are moving. We note, first, as in a previous paragraph, that the improvement in morphology is generic as well as specific. Anatomical analysis is being carried to its limits in a number of directions. The investigations that are connected with the study of cells afford a conspicuous illustration of this fact. Studies in cell-lineage have led to an exact determination of cell-succession in the development of certain animals, and such studies are still in progress. Great progress also has been made in the study of physical structure of living matter. The tracing of cell-lineage is a feat of remarkably accurate and patient work. But, however much this may command our admiration, it has been surpassed (as related in Chapter XI) by investigations regarding the organization of the egg and the analysis of chromosomes. Boveri, Conklin, Wilson, and others have shown that there are recognizable areas within the protoplasm of the egg that have a definite historical relationship to certain structures in process of development. This is the basis upon which rests the doctrine of pre-localization of tissue-forming substances within the protoplasm of the egg.

Anatomy of the Nervous System.—In another direction the progress of anatomical studies is very evident, that is, investigations of the nervous system and the sense-organs. The wonderfully complicated relations of nerve elements have been worked out bv Ramon y Cajal. The studies of Hodge and others upon optical changes occurring within the cells of the nervous system owing to their functional activity have opened a great field for investigation. The studies of Strong, Herrick, and others upon the distribution of nerve-components in the nerves of the head and the investigations of Harrison on the growth and the regeneration of nerve-fibers give illustrations of current tendencies in biological investigation. The analvsis of the central nervous system into segmental divisions on the basis of functional activity (Johnston) is still another illustration.

The Application of Biological Facts to the Benefit of Mankind.—The practical application of biology to the benefit of mankind is a striking feature of present-day tendencies. The activity set on foot by the researches of Pasteur, Koch, and others has created a department of technical biology of the greatest importance to the human race.

Under the general heading should be included tne demonstration of the connection between insects and the propagation of yellow fever, malaria, and other disorders; and as an illustration of activity in 1907, we think of the commission recently appointed to investigate the terrible scourge of the sleeping-sickness which has been prevalent in Africa. Here also we would group studies of a pathological character on blood-immunitv, toxin and antitoxin, also studies on the inoculation for the prevention of various diseases that affect animals and mankind. Very much benefit has alreadv accrued from the practical application of biological researches of this nature, which, in reality, are still in their infancy.

We find the application of biological facts to agriculture

in the form of soil-inoculation, in the tracing of the sources of nitrates in the soil, and studies of the insects injurious to vegetation; their further application to practical forestry and in sanitary sciences. This kind of research is also applied to the study of food-supply for fishes, as in the case of Plankton studies.

The Establishment and Maintenance of Biological Laboratories.—The establishment of seaside biological observatories and various other stations for research have had a great influence on the development of biology. The most famous biological station is that founded at Naples (Fig. 123) in 1872 by Anton Dohrn, and it is a gratification to biologists to know that he still remains its director. This international station for research has stimulated, and is at present stimulating, the growth of biology by providing the best conditions for carrying on researches and by the distribution of material which has been put up at the seacoast by the most skilled preservators. There are many stations modeled after that at Naples. The Marine Biological Laboratory at Woods Holl, Mass., is of especial prominence, and the recently reorganized Wistar Institute of Anatomy at Philadelphia is making a feature of the promotion of anatomical researches, especially those connected with the anatomy of the nervous system.

Laboratories similar to those at the seaside have been established on several fresh-water lakes. The studies carried on in those places of the complete biology of lakes, taking into account the entire surroundings of organisms, are very interesting and important.

Under this general head should be mentioned stations under the control of the Carnegie Institution, the various scientific surveys under the Government, and the United States Fish Commission, which carries on investigations in the biology of fishes as well as observations that affect their use

F G. 123. The Biologica Station at Naples.

as articles of diet. The combined output of the various laboratories and stations of this nature is very considerable, and their influence upon the progress of biology is properly included under the head of present tendencies.

The organization of laboratories in our great universities and their product exercise a wide influence on the progress of biology, that science having within twenty-five years come to occupy a position of great importance among the subjects of general education.

Establishment and Maintenance of Technical Periodicals. —It is manifestly very important to provide means for the publication of results and, as needed, to have technical periodicals established and properly maintained. Their maintenance can not be effected on a purely commercial basis, and the result is that some of our best periodicals require financial assistance in order to exist at all. The subsidizing and support of these periodicals aid materially in the biological advance. A typical technical periodical is Schultze's famous *Archiv für Mikroscopische Anatomie*, founded in 1864 by Schultze and continued to the present time. Into its pages go the highest grade of investigations, and its continued existence has a salutary influence upon the progress of biology. The list of technical periodicals would be too long to name, but among others the *Morphologisches Jahrbuch* of Gegenbaur, and Koelliker's *Zeitschrift für Wissenschaftliche Zoologie* have had wide influence. In England the *Quarterly Journal of Microscopical Science* is devoted to morphological investigations, while physiology is provided for in other journals, as it is also in Germany and other countries. In the United States the *Journal of Morphology*, edited by C. O. Whitman, passed through seventeen volumes and was maintained on the highest plane of scholarship. The fine execution of the plates and the high grade of typographical work made this journal conspicuous. It represents in every way an enterprise of which Americans can be justly proud. The *American Journal of Anatomy* is now filling the field left unoccupied by the cessation of the *Journal of Morphology*.* In the department of experimental work many journals have sprung up, as *Biometrica*, edited by Carl Pearson, Roux's *Archiv für Entwicklungs-Mechanik*, the *Journal of Experimental Zoology* recently established in the United States, etc., etc.

Exploration of the Fossil Records.—Explorations of the fossil records have been recently carried out on a scale never before attempted, involving the expenditure of large sums, but bringing results of great importance. The American Museum of Natural History, in New York City, has carried on an extensive survey, which has enriched it with wonderful collections of fossil animals. Besides explorations of the fossil-bearing rocks of the Western States and Territories, operations in another locality of great importance are conducted in the Fayûm district of Egypt. The result of the studies of these fossil animals is to make us acquainted not only with the forms of ancient life, but with the actual line of ancestry of many living animals. The advances in this direction are most interesting and most important. This extensive investigation of the fossil records is one of the present tendencies in biology.

Conclusion.—In brief, the chief tendencies in current biological researches are mainly included under the following headings: Experimental studies in heredity, evolution, and animal behavior; more exact anatomical investigations, especially in cytology and neurology, the promotion and dissemination of knowledge through biological periodicals; the provision of better facilities in specially equipped laboratories, in the

*It is a source of gratification to biologists that—thanks to the Wistar Institute of Anatomy—the publication of the *Journal of Morphology* is to be continued.

application of results to the benefit of mankind, and in the investigation of the fossil records.

The atmosphere of thought engendered by the progress of biology is beneficial in every way. While its progress has dealt the death-blow to many superstitions and changed materially views regarding the universe, it is gratifying to think that it has not been iconoclastic in its influence, but that it has substituted something better for that which was taken away. It has given a broader and more wholesome basis for religion and theories of ethics; it has taught greater respect for truth and morality. However beneficial this progress has been in the past, who can doubt that the mission of biology to the twentieth century will be more important than to the past, and that there will be embraced in its progress greater benefits than any we have yet known?

READING LIST

The books and articles relating to the history of biology are numerous. Those designated below embrace some of the more readily accessible ones. While some attention has been given to selecting the best sources, no attempt has been made to give a comprehensive list.

I. GENERAL REFERENCES

CUVIER. Histoire des Sciences Naturelles. 5 vols., 1841–1845. Excellent. Written from examination of the original documents.

CARUS. Geschichte der Zoologie, 1872. Also Histoire de la Zoologie, 1880. A work of scholarship. Contains excellent account of the Physiologus.

SACHS. History of Botany, 1890. Excellent. Articles in the *Botanical Gazette* for 1895 supplement his account by giving the more recent development of botany.

WHITE. A History of the Warfare of Science with Theology in Christendom, 2 vols., 1900. Good account of Vesalius and the overthrow of authority in science.

WHEWELL. History of the Inductive Sciences, vol. II, 1863. Lacks insight into the nature of biology and the steps in its progress. Mentioned because so generally known.

WILLIAMS. A History of Science, 5 vols., 1904. Finely illustrated. Contains many defects in the biological part as to the relative rank of the founders: Vesalius diminished, Paracelsus magnified, etc. Also, the Story of Nineteenth Century Science, 1900. Collected articles from *Harper's Magazine*. Good portraits. Uncritical on biological matters.

THOMSON. The Science of Life, 1899. An excellent brief history of biology.

FOSTER. Lectures on the History of Physiology, 1901. Fascinatingly written. Notable for poise and correct estimates, based on the use of the original documents.

GEDDES. A Synthetic Outline of the History of Biology. *Proc. Roy. Soc. Edinb.*, 1885–1886. Good.

RICHARDSON. Disciples of Æsculapius, 2 vols., 1901. Collected papers from *The Asclepiad.* Sympathetic accounts of Vesalius, Malpighi, J. Hunter, and others. Good illustrations.

LANKESTER. The History and Scope of Zoology, in The Advancement of Science, 1890. Good. Same article in Ency. Brit. under the title of Zoology.

SPENCER. Principles of Biology, 2 vols., 1866.

HERTWIG. The Growth of Biology in the Nineteenth Centurv, *Ann. Rept. Smithson. Inst.*, 1900.

BUCKLE. History of Civilization, vol. I, second edition, 1870.

MACGILIVRAY. Lives of Eminent Zoölogists from Aristotle to Linnæus.

MERZ. A History of European Thought in the Nineteenth Century, vol. II, Scientific Thought, 1903.

ROUTLEDGE. A Popular History of Science. General and uncritical as to biology.

HOEFER. Histoire de la Zoologie, 1873. Not very good.

ENCYCLOPÆDIA BRITANNICA. Among the more excellent articles are: Biology by Huxley; Protoplasm by Geddes; History of Anatomy by Turner.

CHAMBERS'S ENCYCLOPÆDIA. New Edition. Discerning articles by Thomson on the Cell-theory, by Geddes on Biology, Evolution.

NOUVELLE BIOGRAPHIE GÉNÉRALE. Good articles on the older writers. Often unreliable as to dates.

HAECKEL. The historical chapters in The Evolution of Man, 1892, and Anthropogenie, fifth edition, 1903. Good.

HAECKEL. The History of Creation, vol. I, 1884.

HERTWIG. The General Survey of the History of Zoölogy in his Manual of Zoölogy, 1902. Brief but excellent.

PARKER and HASWELL. Text-book of Zoölogy, 1897. Historical chapter in vol. II.

NICHOLSON. Natural Historv, its Rise and Progress in Britain, 1886. Also Biology.

PETTIGREW. Gallery of Medical Portraits, 5 vols. Contains many portraits and biographical sketches of men of general influence, as Bichat, Galen, Malpighi, etc.

PUSCHMANN. Handbuch der Geschichte der Medizin, 3 vols. Good for topics in anatomy and physiology.

BAAS. The History of Medicine, 1889.

RADL. Geschichte der Biologischen Theorien seit dem Ende des Siebzehnten Jahrhundert, 1905.

JANUS. A Periodical devoted to the history of medicine and natural science, founded in 1896.

ZOOLOGISCHE ANNALEN. Founded by Max Braun in 1904 in the interests of the history of zoölogy.

MITTEILUNGEN ZUR GESCHICHTE DER MEDIZIN UND NATURWISSENSCHAFTEN, founded 1901.

SURGEON GENERAL'S LIBRARY. The Catalogue should be consulted for its many biographical references to biologists. The Library is especially rich in historical documents, as old anatomies, physiologies, zoölogies, etc.

EVOLUTION. The bibliography of Evolution is given below under the chapters dealing with the evolution theory.

II. SPECIAL REFERENCES

CHAPTER I

ANCIENT BIOLOGICAL SCIENCE: Carus; Botany after 1530, Sachs. ARISTOTLE: Cuvier, a panegyric; Lewes, Aristotle—A Chapter from the History of Science, 1864, a critical study; Huxley, On some Mistakes Attributed to Aristotle; Macgilivray; Aristotle's History of Animals translated in Bohn's Classical Library, 1887. PLINY: Magilivray; Thorndike, The Place of Magic in the Intellectual History of Europe, 1905, chap. III. THE RENAISSANCE: Symonds. EPOCHS IN BIOLOGICAL HISTORY: Geddes (see General List).

CHAPTER II

VESALIUS: Roth, Andreas Vesalius Bruxellensis, the edition of 1892, the standard source of knowledge of Vesalius and his times, contains bibliography, references to his different portraits, the resurrection bone, etc., etc.; Foster (see General List), Lecture I, excellent; Richardson in Disciples of Æsculapius, vol. I, contains pictures, his signature, etc.; Pettigrew; White vol. II, pp. 51–55; The Practitioner, 1896, vol. 56; The Asclepiad, 1885, vol. II; De Humani Corporis Fabrica, editions of 1543 and 1555; Opera Omnia, edited by Boerhaave, 2 vols., 1725. GALEN: Pettigrew; Huxley in his essay on William Harvey.

CHAPTER III

HARVEY: Foster, Lecture II, with quotations excellent; Dalton, History of the Circulation; Huxley, William Harvey, a critical essay; Harvey's Works translated by Willis, with biography, Sydenham Society, 1847; Life of Harvey by D'Arcy Power, 1898; Brooks, Harvey as Embryologist, Bull. Johns Hop. Hospit., vol. VIII, 1897, good. An Anatomical Dissertation upon the Movement of the Heart and Blood in Animals, a facsimile reproduction of the first edition of the famous De Motu Cordis et Sanguinis, 1628. Privately reproduced by Dr. Moreton in 1894. Very interesting.

CHAPTER IV

HOOKE: Biography in encyclopædias, his microscope in Carpenter, The Microscope and Its Revelations, 8th ed., 1900.

MALPIGHI: Richardson, vol. II; Same article in *The Asclepiad*, vol. X, 1893; Atti, Life and Work, in Italian, 1847, portrait; Pettigrew, vol. II; Marcello Malpighi e l'Opera Sua, 1897, a collection of addresses at the unveiling of Malpighi's monument at Crevalcuore, that by Koelliker excellent; Locy, Malpighi, Swammerdam, and Leeuwenhoek, *Pop. Sci. Mo.*, 1901—protrait and pictures from his works; MacCallum, *J. Hop. Univ. Hospit. Bull.* MALPIGHI'S WRITINGS: Opera Omnia, difficult to obtain, the Robt. Littlebury edition, Lond., 1687, contains posthumous papers and biography; separate works not uncommon; Traité du Ver à Soie, Montpellier, 1878, contains his life and works.

SWAMMERDAM: Life by Boerhaave in Biblia Naturæ, 1735; also Bibel der Natur, 1752; also The Book of Nature, 1758; Von Baer, Johann Swammerdam's Leben und Verdienste um die Wissenschaft, 1864, in *Reden*, vol. I; Locy, *loc. cit.*—portrait.

LEEUWENHOEK: New biographical facts in Richardson, vol. I, p. 108; same article in *The Asclepiad*, vol. II, 1885, portrait, signature, and other illustrations; Arcana Naturæ; Selected works in English, 1758; Locy, *Pop. Sci. Mo.*, April, 1901.

CHAPTER V

LYONET: *The Gentleman's Magazine*, LIX, 1789; the famous Traité Anatomique, etc., 1750, 1752, not rare. RÉAUMUR: Portrait and life in *Les Savants Modernes*, p. 332. ROESEL: Portrait and biography in *Der monatlich herausgegebenen Insecten Belustigung*, part IV, 1761; Zeigler in *Natur und Haus*, 1904—nine figs. STRAUS-DÜRCKHEIM: his monograph on Anatomy of the Cockchafer, rather rare. THE MINUTE ANATOMISTS: Straus-Dürckheim, Dufour, Newport, Leidig, etc., in Miall and Denney's The Cockroach, 1886.

DISCOVERY OF THE PROTOZOA: Leeuwenhoek, Müller, Ehrenberg, Dujardin, etc., Kent's Manual of the Infusoria, vol. I. EHRENBERG: Life by Laue, 1895.

CHAPTER VI

THE PHYSIOLOGUS: Carus, White (for titles see General List). GESNER: Brooks in *Pop. Sci. Mo.*, 1885—illustrations; Cuvier, *loc. cit.;* Jardine's Naturalist's Library, vol. VI; Gesner's Historia Animalium, 1551–1585. ALDROVANDI: Naturalist's Library, vol. III; Macgilivray, *loc. cit.* JONSTON: Macgilivray. RAY: Macgilivray; Nicholson; Memorial of, in the Ray Society, 1846; Correspondence of, Ray Soc., 1848. LINNÆUS: Macgilivray; *Janus*, vol. 8, 1903; Cuvier, *loc. cit.;* Agassiz, Essay on Classification, 1859; Jubilee at Upsala, *Science*, Apl. 26, 1907; Caddy, Through the Fields with Linnæus, 1887; The Systema Naturæ, especially the tenth edition, 1758. LEUCKART: Archives de Parasit., vol. I, no. 2; *Nature*, 1898. GENERAL BIOLOGICAL PROGRESS FROM LINNÆUS TO DARWIN: Geddes, Proc. Roy. Soc. Edinb., vol. 13, 1884–1886.

CHAPTER VII

CAMPER: Naturalist's Library, vol. VII; Vorlesungen, by his son, with short sketch of his life, 1793; Cuvier, *loc. cit.; Kleinere Schriften*, 2 vols. with copper plates illustrating brain and ear of fishes, etc., 1782–1785. JOHN HUNTER: The Scientific Works of, 2 vols., 1861; The *Asclepiad*, vol. VIII, 1891; the same article with illustrations in Richardson, *loc. cit.;* Pettigrew, *loc. cit.* VICQ D'AZYR: Cuvier, *loc. cit.;* Huxley in Life of Owen, p. 289; His works in 6 vols., 1805. CUVIER: Life by Flourens; Memoirs by Mrs. Lee, 1833; Buckle, Hist. Civ., vol. I, p. 633 et seq.; Lettres de Geo. Cuvier à C. M. Paff, 1788–1792, translated from the German, 1858. Cuvier's numerous writings—The Animal Kingdom, Leçons d'Anat. Comparée, etc. —are readily accessible. H. MILNE-EDWARDS: Biographical sketch in *Ann. Rept. Smithson. Inst.* for 1893. LACAZE-DUTHIERS: Life with portraits in *Archives de Zool. Expériment.*, vol. 10, 1902. RICHARD OWEN: Life and Letters, 2 vols., 1894; Clark, Old Friends at Cambridge and Elsewhere, p. 349 et seq. J. FR. MECKEL: Carus, *loc. cit.* GEGENBAUR: Erlebtes und Erstrebtes, portrait, 1901; Anat. Anz., vol. 23, 1903; *Ann. Rept. Smithson. Inst.*, 1904. COPE: Osborn in *The Century*, vol. 33, 1897; Gill, Edward Drinker Cope, Naturalist, A Chapter in the History of Science, *Am. Naturalist*, 1897; Obituary notice, with portraits, *Am. Naturalist*, 1897; *Pop. Sci. Mo.*, vol. 19, 1881.

CHAPTER VIII

BICHAT: Pettigrew; Buckle, Hist. Civ., vol. I, p. 639; The Hundred Greatest Men; *Les Savants Modernes*, p. 394; *The Practitioner*, vol. 56, 1896. KOELLIKER: His Autobiography, Erinnerungen aus Meinem

Leben, 1899, several portraits, interesting; Weldon, Life and Works in *Nature*, vol. 58, with fine portrait; Sterling, *Ann. Rept. Smithson. Inst.*, 1905. SCHULTZE: Portrait and Necrology by Schwalbe in *Archiv für Mikroscop. Anat.*, vol. 10, 1874; See further under chapter XII. VIRCHOW: *J. Hop Univ. Circulars*, vol. XI, 1891, Celebration of Seventieth Birthday of Virchow, Addresses by Osler, Welch, and others; Jacobi, *Medical Record, N. Y.*, vol. XX, 1881, good; Israel, in *Ann. Rept. Smithson. Inst.*, 1902. LEYDIG: Brief sketch in his Horæ Zoologicæ, 1902. RAMON Y CAJAL: Portrait in Tenth Anniversary of Clark University, 1899.

CHAPTER IX

The best brief account cf the Rise of Physiology in Verworn's General Physiology, 1899. More recent German editions of the same work. Historical outline in Rutherford's Text-Book of Physiology, 1880. GALEN'S PHYSIOLOGY: Verworn. HARVEY: See references under Chapter III; The analysis of his writings by Willis in The Works of Harvey, translated into English, Sydenham Soc., 1847; See also Dr. Moreton's facsimile reproduction of the first edition (1628) of De Motu Cordis et Sanguinis, 1894. HALLER: Fine portrait in his Elementa Phvsiologiæ, 1758; English translations of the Elementa. CHARLES BELL: Pettigrew; Good summary in Foster's Life of Claude Bernard, p. 38 et seq. JOHANNES MÜLLER: His life, complete list of works, etc., in Gedächtnissrede auf Johannes Müller by Du Bois-Reymond, 1860; *Eloge* by Vichow in *Edinburgh Med. Journ.*, vol. 4; Picture of his monument in Coblenz, *Archiv f. Mik. Anat.*, vol. 55; Briefe von J. Müller an Andres Retzius (1830–1857), 1900; His famous Handbuch der Physiologie and English translations should be inspected. LUDWIG: Burdon-Sanderson, Ludwig and Modern Physiology, *Sci. Progress*, vol. V, 1896; The same article in *Ann. Rept. Smithson. Inst.*, 1896. CLAUDE BERNARD: Life by M. Foster, 1899, excellent.

CHAPTER X

Good general account of the Rise of Embryology in Koelliker's Embryologie, 1880; Minot, Embryology and Medical Progress, *Pop. Sci. Mo.*, vol. 69, 1906; Eycleshymer, A Sketch of the Past and Future of Embryology, *St. Louis Med. Rev.*, 1904. HARVEY: As Embryologist, Brooks in *J. Hop. Univ. Hospit. Bull.*, vol. VIII, 1897. See above, Chaps. III and IX for further references to Harvey. MALPIGHI: in Embryology, Locy in *Pop. Sci. Mo.*, 1905—portrait and selected sketches from his embryological treatises. WOLFE: Wheeler, Wolff and the Theoria Generationis, in Woods Holl Biological Lectures, 1898; Kirchoff in *Jenaische Zeitschr.*, vol. 4, 1868; Waldeyer, Festrede in Sitzbr. d. K. Preus. Akad. d. Wissenschaft., 1904; Haeckel in Evolution of Man, vol. I, 1892. BONNET AND PRE-DELINEATION: Whitman, Bonnet's Theory of Evolution, also Evolution and Epigenesis, both in Woods Holl Biological Lectures, 1895. VON BAER: Leben und Schriften, his autobiography (1864), 2d edition, 1886; Life by Steida, 1886; Obituary, *Proc. Roy. Soc.*, 1878; Waldeyer in *Allg. Wien. Med. Ztg.*, 1877; *Nature*, vol. 15; Life by Stölzle, 1897; Haeckel, *loc. cit.*, vol. I; Locy, V. Baer and the Rise of Embryology, *Pop. Sci. Mo.*, 1905; Fine portrait as young man in *Harper's Mag.* for 1899; *Rev. Scient.*, 1879. KOWALEVSKY: Lankester in *Nature*, vol. 66, 1902; Portrait and biog. in *Ann. Mus. Hist. Nat. Marseille*, vol. 8, 1903. BALFOUR: M. Foster in *Nature*, vol. 29, 1882; Also Life with portrait in the Memorial Edition of Balfour's Works; Waldeyer in *Arch. f. Mik. Anat.*, vol. 21, 1882; Osborn Recollections, with portrait, *Science*, vol. 2, 1883. HIS: Mall in *Am. Journ. Anat.*, vol. 4, 1905; Biography in *Anat. Anz.*, vol. 26, 1904.

CHAPTER XI

The Cell-Doctrine by Tyson, 1878. THE CELL-THEORY, Huxley, *Medico-chir. Review*, 1853, also in Scientific Memoirs, vol. I, 1898; The Modern Cell-Theory, M'Kendrick, *Proc. Phil. Soc. Glasgow*, vol. XIX, 1887; The Cell-Theory, Past and Present, Turner, *Nature*, vol. 43, 1890; The Cell-Doctrine, Burnett, *Trans. Am. Med. Assn.*, vol. VI, 1853; First illustration of cells in Rob't Hooke's Micrographia, 1665, 1780, etc.; The Cell in Development and Inheritance, Wilson, 1896; Article Cell, in Chambers's (New) Cyclopædia, by Thomson. SCHLEIDEN: Sketch of, *Pop. Sci. Mo.*, vol. 22, 1882–1883; Sachs' Hist. of Botany 1890; Translation of his original paper of 1838 (Ueber Phytogenesis)—illustrations—Sydenham Soc., 1874. SCHWANN: Life, *Pop. Sci. Mo.*, vol. 37, 1900; Sa Vie et Ses Travaux Frédéricq, 1884; Nachruf, Henle, *Archiv f. Mik. Anat.*, vol. 21, 1882; Lankester, *Nature*, vol. XXV, 1882; *The Practitioner*, vol. 49, 1897; *The Catholic World*, vol. 71, 1900. Translation of his contribution of 1839 (Mikroscopische Untersuchungen ueber die Uebereinstimmung in der Structur und dem Wachstum der Thiere und Pflanzen), Sydenham Soc., 1847.

CHAPTER XII

On the Physical Basis of Life, Huxley, 1868; Reprint in Methods and Results, 1894. Article Protoplasm in Ency. Brit. by Geddes. DUJARDIN: *Notice Biographique*, with portraits and other illustrations, Joubin, *Archives de Parasitol.*, vol. 4, 1901; portrait of Dujardin hitherto unpublished. Dujardin's original description of Sarcode, *Ann. des Sci. Nat.* (*Botanique*), vol. 4, p. 367, 1835. VON MOHL: Sachs' History of Botany, 1890. Translation of his researches, Sydenham Soc., 1847. COHN: Blätter der Erinnerung, 1898, with portrait. SCHULTZE: Necrology, by Schwalbe in *Archiv f. Mik. Anat.*, vol. 10, 1874, with portrait. Schultze's paper founding the protoplasm doctrine in *Archiv f. Anat. und Phys.*, 1861, entitled Ueber Muskelkörperchen und das was man eine Zelle zu nennen habe.

CHAPTER XIII

SPONTANEOUS GENERATION: Tyndall, *Pop. Sci. Mo.*, vol. 12, 1878; Also in Floating Matter of the Air, 1881; J. C. Dalton in *N. Y. Med. Journ.*, 1872; Dunster, good account in *Proc. Ann Arbor Sci. Assn.*, 1876; Huxley, *Rept. Brit. Assn. for Adv. Sci.*, 1870, republished in many journals, reprint in Scientif. Memoirs, vol. IV, 1901. REDI: Works in 9 vols., 1809–1811, with life and letters and portraits; Good biographical sketch in *Archives de Parasitol.*, vol. I, 1898; Redi's Esperienze Intorno Alla Generazione Degl'Insetti, 2 plates, first edition, 1668, in Florence, 40; reprinted at various dates, not uncommon. SPALLANZANI: Foster, Lects. on Physiol.; Huxley, *loc. cit.;* Dunster, *loc. cit.;* L'Abbato Spallanzani, by Pavesi, 1901, portrait. POUCHET: His treatise of historical importance—Hétérogénie; ou Traité de la Génération Spontanée, basé sur des Nouvelles Expériences, 1859. PASTEUR: Life by René Vallery-Radot, 2 vols., 1902; Percy and G. Frankland, 1901; Pasteur at Home, illustrated, Tarbell in *McClure's Mag.*, vol. I, 1893; Also *McClure's*, vol. 19, 1902, review of Vallery-Radot's Life of Pasteur; *Nature*, vol. 52, 1895; *Les Savants Modernes*, p. 316; Life by his son-in-law, translated by Lady Hamilton, 1886; Sketches of Pasteur, very numerous. BACTERIOLOGY: Woodhead, Bacteria and their Products, 1891; Fraenkel, Text-Book of Bacteriology, 1891; Prudden, The Story of Bacteria, etc., 1891. GERM-THEORY OF DISEASE: Crookshank's Bacteriology, 3d edition, 1890. KOCH: *Pop. Sci. Mo.*, vol. 36, 1889; *Review of Reviews*, vol. 2, 1890; Sketches and references to his discoveries numerous. LISTER: *Pop. Sci. Mo.*, vol. 52, 1898; *Review of Reviews*, vol. 14, 1896; celebration of Lister's 80th birthday, *Pop. Sci. Mo.*, June, 1907; *Janus*, vol. 5, 1900. The New Microbe Inoculation of Wright, *Harper's Mag.*, July, 1907.

CHAPTER XIV

The History and Theory of Heredity, J. A. Thomson, *Proc. Roy. Soc. Edinb.*, vol. XVI, 1889; Chapter on Heredity in Thomson's Science of Life, 1899; also in his Study of Animal Life, 1892. MENDEL: Mendel's Principles of Heredity, with translations of his original papers on hybridization, Bateson, 1902; Mendel's Versuche über Pflanzenhybriden, two papers (1865 and 1869), edited by Tschermak, 1901; *Ann. Rept. Smithson. Inst.*, 1901–1902; *Pop. Sci. Mo.*, vol. 62, 1903; vol. 63, 1904; *Science*, vol. 23, 1903. GALTON: *Pop. Sci. Mo.*, vol. 29, 1886; *Nature*, vol. 70, 1907; Galton's Natural Inheritance, 1889. WEISMANN: Brief Autobiography, with portrait, in *The Lamp*, vol. 26, 1903; Solomonsen, Bericht über die Feier des 70 Geburtstages von August Weismann, 1904; Weismann's The Germ-Plasm, 1893, and The Evolution Theory, 1904.

CHAPTER XV

History of Geology and Paleontology, Zittel, 1901. The Founders of Geology, Geikie, 2d edition, 1905. History and Methods of Paleontological Discovery, Marsh, *Proceed. Am. Adv. Sci.*, 1879. Same article in *Pop. Sci. Mo.*, vol. 16, 1879–1880. The Rise and Progress of Paleontology, Huxley, *Pop. Sci. Mo.*, vol. 20, 1882. LYELL: Charles Lyell and Modern Geology, Bonney, 1895; Sketch in *Pop. Sci. Mo.*, vol. I, 1872, also vol. 20, 1881–1882. OWEN: Life of, by his grandson, 2 vols., 1894; See also above under Chapter VII. AGASSIZ: Life and Correspondence, by his wife, 2 vols., 1885; Life, letters and works, Marcou, 2 vols., 1896; What we Owe to Agassiz, Wilder, *Pop. Sci. Mo.*, July, 1907; Agassiz at Penikese, *Am. Nat.*, 1898. COPE: A Great Naturalist, Osborn in *The Century*, 1897; See above, under Chapter VII, for further references. MARSH: *Pop. Sci. Mo.*, vol. 13, 1878; Sketches of, *Nature*, vol. 59, 1898–99; *Science*, vol. 9, 1899; *Am. J. Sci.*, vol. 157, 1899. ZITTEL: Biographical Sketch with portrait, Schuchert, *Ann. Rept. Smithson. Inst.*, 1903–1904. Osborn, Papers on Paleontological Discovery in *Science* from 1899 onward. The Favûm Expedition of the Am. Museum of Nat. History, *Science*, March 29, 1907.

NOTE. Since the four succeeding chapters deal with the Evolution Theory, it may be worth while to make a few general comments on the literature pertaining to Organic Evolution. The number of books and articles is very extensive, and I have undertaken to sift from the great number a limited list of the more meritorious. Owing to the prevalent vagueness regarding evolution theories, one is likely to read only about Darwin and Darwinism. This should be avoided by reading as a minimum some good reference on Lamarck, Weismann, and De Vries, as well as on Darwin. It is well enough to begin with Darwin's Theory, but it is not best to take his Origin of Species as the first book. To do this is to place oneself fifty years in the past. The evidences of Organic Evolution have greatly multiplied since 1859, and a better conception of Darwin's Theory can be obtained by reading first Romanes's Darwin and After Darwin, vol. I. This to be followed by Wallace's Darwinism, and, thereafter, the Origin of Species may be taken up. These will give a good conception of Darwin's Theory, and they should be followed by reading in the order named: Packard's Lamarck; Weismann's The Evolution Theory; and De Vries's The Origin of Species and Varieties by Mutation. Simultaneously one may read with great profit Osborn's From the Greeks to Darwin.

CHAPTER XVI

GENERAL: Romanes, Darwin and After Darwin, 1892, vol. I, chaps. I–V; Same author, The Scientific Evidences of Organic Evolution; Weismann Introduction to the Evolution Theory, 1904; Osborn, Alte und Neue Probleme der Phylogenese, *Ergebnisse der Anat. u. Entwickel.*, vol. III, 1893; Ziegler, Ueber den derzeitigen Stand der Descendenzlehre in der Zoologie, 1902; Jordan and Kellogg, Evolution and Animal Life, 1907, chaps. I and XIV. EVOLUTIONARY SERIES—SHELLS: Romances, *loc. cit.;* Hyatt, Transformations of Planorbis at Steinheim, *Proc. Am. Ass. Adv. Sci.*, vol. 29, 1880. HORSE: Lucas, The Ancestry of the Horse, *McClure's Mag.*, Oct., 1900; Huxley, Three Lectures on Evolution, in Amer. Addresses. EMBRYOLOGY—RECAPITULATION THEORY: Marshall, Biolog. Lectures and Addresses, 1897; Vertebrate Embryology, 1892; Haeckel, Evolution of Man, 1892. PRIMITIVE MAN: Osborn, Discovery of a Supposed Primitive Race of Men in Nebraska, *Century Mag.*, Jan., 1907; Haeckel, The Last Link, 1898. Huxley, Man's Place in Nature, collected essays, 1900; published in many forms. Romanes, Mental Evolution in Man and Animals.

CHAPTER XVII

LAMARCK: Packard, Lamarck, the Founder of Evolution, His Life and Work, with Translations of his Writings on Organic Evolution, 1901; Lamarck's Philosophie Zoologique, 1809. Recherches sur l'Organisation des corps vivans, 1802, contains an early, not however the first statement of Lamarck's views. For the first published account of Lamarck's theory see the introduction to his Système des Animaux sans Vertèbres, 1801. NEO-LAMARCKISM: Packard, *loc. cit.;* also in the Introduction to the Standard Natural History, 1885; Spencer, The Principles of Biology, 1866 —based on the Lamarckian principle. Cope, The Origin of Genera, 1866; Origin of the Fittest, 1887; Primary Factors of Organic Evolution, 1896, the latter a very notable book. Hyatt, Jurassic Ammonites, *Proced. Bost. Sci. Nat. Hist.*, 1874. Osborn, *Trans. Am. Phil. Soc.*, vol. 16, 1890. Eigenmann, The Eves of the Blind Vertebrates of North America, *Archiv f. Entwickelungsmechanik*, vol. 8, 1899.

DARWIN'S THEORY (For biographical references to Darwin see below under Chapter XIX): Wallace, Darwinism, 1889; Romanes, Darwin and After Darwin, vol. I, 1892; Metcalf, An Outline of the Theory of Organic Evolution, 1904, good for illustrations. COLOR: Poulton, The Colors of Animals; Chapters in Weismann's The Evolution Theory, 1904. MIMICRY: Weismann, *loc. cit.* SEXUAL SELECTION: Darwin, The Descent of Man, new ed., 1892. INADEQUACY OF NAT. SELECTION: Spencer, The Inadequacy of Natural Selection, 1893; Morgan, Evolution and Adaptation, 1903. Kellogg, Darwinism To-day, 1907, contains a good account of criticisms against Darwinism.

CHAPTER XVIII

Weismann's The Evolution Theory, translated by J. A. and Margaret Thomson, 2 vols., 1904, contains the best statement of Weismann's views. It is remarkably clear in its exposition of a complicated theory. The Germ-Plasm, 1893; Romanes's An Examination of Weismannism, 1893. INHERITANCE OF ACQUIRED CHARACTERS: Weismann's discussion, *loc. cit.*, vol. II, very good. Romanes's Darwin and After Darwin, vol. II. PERSONALITY OF WEISMANN: Sketch and brief autobiography, in *The Lamp*, vol. 26, 1903, portrait; Solomonsen, Bericht über die Feier des 70 Geburtstages von August Weismann, 1905, 2 portraits.

MUTATION-THEORY OF DE VRIES: Die Mutations-Theorie, 1901; Species and Varieties, their Origin by Mutation, 1905; Morgan, Evolution and Adaptation, 1903, gives a good statement of the Mutation Theory, which is favored by the author; Whitman, The Problem of the Origin of Species, *Congress of Arts and Science, Universal Exposition, St. Louis*, 1904; Davenport, Evolution without Mutation, *Journ. Exp. Zool.*, April, 1905.

CHAPTER XIX

For early phases of Evolutionary thought consult Osborn, From the Greeks to Darwin, 1894, and Clodd, Pioneers of Evolution, 1897. SUAREZ AND THE DOCTRINE OF SPECIAL CREATION: Huxley, in Mr. Darwin's Critics, *Cont. Rev.*, p. 187, reprinted in Critiques and Addresses, 1873. BUFFON: In Packard's Life of Lamarck, chapter 13. E. DARWIN: Krause's Life of E. Darwin translated into English, 1879; Packard, *loc. cit.* GOETHE: Die Idee der Pflanzenmetamorphose bei Wolff und bei Goethe, Kirchoff, 1867; Goethe's Die Metamorphose der Pflanzen, 1790. OKEN: His Elements of Physiophilosophy, Ray Soc., 1847. CUVIER AND ST. HILAIRE: Perrier, La Philosophie Zoologique avant Darwin, 1884; Osborn, *loc. cit.* DARWIN AND WALLACE: The original communications of Darwin and Wallace, with a letter of transmissal signed by Hooker and Lyell, published in the *Trans. Linnæan Soc.* for 1858, were reprinted in the *Pop. Sci. Mo.*, vol. 60, 1901. DARWIN: Personality and biography (For references to his theory see under Chapter XVII); Life and letters by his son, 3 vols., 1887, new ed., 1896; More Letters of Charles Darwin, 2 vols., 1903; Chapter in Marshall's Lectures on the Darwinian Theory; Darwin, Naturalist's Voyage around the World, 1880; Gould, Biographical Clinics, for Darwin's illness due to eye-strain; Poulton, Chas. Darwin and the Theory of Natural Selection, 1896. WALLACE: My Life, 2 vols., 1905; The Critic, Oct., 1905. HUXLEY: Life and Letters by his son, 1901; Numerous sketches at the time of his death, 1895, in *Nature*, *Nineteenth Century*, *Pop. Sci. Mo.*, etc., etc. HAECKEL: His Life and Work by Bölsche, 1906.

CHAPTER XX

It is deemed best to omit the references to Technical papers upon which the summaries of recent tendencies are based. Morgan's Experimental Zoology, 1907. Jennings, Behavior of the Lower Organisms, 1906. Mosquitoes and other insects in connection with the transmission of disease, see Folsom, Entomology, 1906, chapter IX, p. 299. BIOLOGICAL LABORATORIES: Dean, The Marine Biological Stations of Europe, *Ann. Rept. Smithson. Inst.*, 1894; Marine Biolog. Station at Naples, *Harper's Mag.*, 1901; The *Century*, vol. 10 (Emily Nunn Whitman); Williams, A History of Science, vol. V, chapter V, 1904; *Am. Nat.*, vol. 31, 1897; *Pop. Sci. Mo.*, vol. 54, 1899; *ibid.*, vol. 59, 1901. Woods Holl Station—A Marine University, *Ann. Rept. Smithson. Inst.*, 1902.

INDEX

INDEX

I

J

K

L

American Science Series

The two principal objects of the series are to supply authoritative books whose principles are, so far as practicable, illustrated by American facts, and also to supply the lack that the advance of science perennially creates, of text-books which at least do not contradict the latest generalizations.

Physics.

By A. L. KIMBALL, Professor in Amherst College. (*In preparation.*)

Physics.

By GEORGE F. BARKER. x + 902 pp. $3.50.

Chemistry.

By IRA REMSEN, President of the Johns Hopkins University.

Advanced Course. xxii + 853 pp. $3.00.
College Chemistry. xx + 689 pp. $2.25.
Briefer Course. xxiv + 516 pp. $1.25.
Elementary Course. x + 287 pp. 80 cents.

Astronomy.

By SIMON NEWCOMB and EDWARD S. HOLDEN.

Advanced Course. xii + 512 pp. $2.00.
Briefer Course. x + 366 pp. $1.20.
Elementary Course. xv + 446 pp. $1.20.

Geology.

By THOMAS C. CHAMBERLIN and ROLLIN D. SALISBURY, Professors in the University of Chicago. 3 vols. 8vo.

Vol. I. Geological Processes and their Results. xix + 654 pp. $4.00.

Vols. II and III. Earth History. xxxvii + 1316 pp. (*Not sold separately.*) $8.00.

Physiography.

By ROLLIN D. SALISBURY, Professor in Chicago University.

Advanced Course. xx + 770 pp. $3.50.
Briefer Course. viii + 531 pp. $1.50.

with Chapter on Anatomy of Birds by FREDERIC A. LUCAS, Chief Curator Brooklyn Museum of Arts and Sciences, and edited by ROBERT RIDGWAY, Curator of Birds, U. S. National Museum.

REPTILES AND BATRACHIANS, by LEONHARD STEJNEGER, Curator of Reptiles, U. S. National Museum.

Section B. A Shorter Natural History, mainly by the Authors of Section A, preserving its popular character, its proportional treatment, and its authority so far as that can be preserved without its fullness. Size not yet determined.

II. CLASSIFICATION OF NATURE

Section A. Realms of Nature. Detailed treatment of various departments in a literary and popular way. 8vo. 7½x10¼ in.

Already publisht :

FERNS, by CAMPBELL E. WATERS, of Johns Hopkins University. 8vo, pp. xi+362. $3.00 net; by mail, $3.30.

Section B. Identification Books—

1. Library Series, very full descriptions. 8vo. 7½x10¼ in.

Already publisht ·

NORTH AMERICAN TREES, by N. L. BRITTON, Director of the New York Botanical Garden. $7.00 net; carriage extra.

2. Pocket Series, "How to Know," brief and in portable shape.

III. FUNCTIONS OF NATURE

These books will treat of the relation of facts to causes and effects—of heredity in organic Nature, and of the environment in all Nature. 8vo. 6⅝x8⅞ in.

Already publisht :

THE BIRD : ITS FORM AND FUNCTION, by C. W. BEEBE, Curator of Birds in the New York Zoological Park. 8vo, 496 pp. $3.50 net; by mail, $3.80.

Arranged for :

THE INSECT : ITS FORM AND FUNCTION, by VERNON L. KELLOGG, Professor in the Leland Stanford Junior University.

THE FISH : ITS FORM AND FUNCTION, by H. M. SMITH, of the U. S. Bureau of Fisheries.

IV. WORKING WITH NATURE

How to propagate, develop and care for the plants and animals. The volumes in this group cover such a range of subjects that it is impracticable to make them of uniform size.

Already publisht :

NATURE AND HEALTH, by Edward Curtis, Professor Emeritus in the College of Physicians and Surgeons. 12mo. $1.25 net; by mail, $1.37.

Arranged for :

PHOTOGRAPHING NATURE, by E. R. Sanborn, Photographer of the New York Zoological Park.

THE SHELLFISH INDUSTRIES, by James L. Kellogg, Professor in Williams College.

CHEMISTRY OF DAILY LIFE, by Henry P. Talbot, Professor of Chemistry in the Massachusetts Institute of Technology.

DOMESTIC ANIMALS, by William H. Brewer, Professor Emeritus in Yale University.

THE CARE OF TREES IN LAWN, STREET AND PARK, by B. E. Fernow, Professor of Forestry in the University of Toronto.

V. DIVERSIONS FROM NATURE

This division will include a wide range of writings not rigidly systematic or formal, but written only by authorities of standing. Large 12mo. 5¼x8⅛ in.

FISH STORIES, by David Starr Jordan and Charles F. Holder.

HORSE TALK, By William H. Brewer.

BIRD NOTES, by C. W. Beebe.

INSECT STORIES, by Vernon L. Kellogg.

VI. THE PHILOSOPHY OF NATURE

A Series of volumes by President Jordan, of Stanford University, and Professors Brooks of Johns Hopkins, Lull of Yale, Thomson of Aberdeen, Przibram of Austria, zur Strassen of Germany, and others. Edited by Professor Kellogg of Leland Stanford. 12mo. 5⅛x7½ in.

HENRY HOLT AND COMPANY, New York

June, '08.

This book has been produced in line with the EU GPSR guidelines about the safety of products.

The General Product Safety Regulation is the European Union's updated framework for ensuring that all consumer products, including books, are safe for consumers.

This book has been printed by CPI books GmbH. The printer has issued safety certificates for the materials - like ink, paper and glue - being used.

The product identifier is: 9786256629813

The author is responsible for the content of the book, is the publisher of the works and bears full responsibility for it.

The book has been published via Bookmundo. Bookmundo enables any author to share their stories with the rest of the world via printed books and ebooks and a broad distribution network.

Bookmundo will act as an intermediary in regard to questions about safety and will address them to the printer / author. Should there be any question in regard to the safety of the product, please contact us.

Bookmundo
Delftsestraat 33
3013AE Rotterdam
The Netherlands
info@bookmundo.com

Kolibri 360 GmbH
Ferdinand-Jühlke-Straße 7
99095 Erfurt, Deutschland
produktsicherheit@kolibri360.de